catching Bluegill

**Proven Methods
No Matter How
You Fish**

John Tertuliani 2008

catching Bluegill

**Proven Methods
No Matter How
You Fish**

JOHN TERTULIANI

Lotic Books

Lotic Books
P.O. Box 543
Hilliard, Ohio 43026

Copyright © 2006 by John Tertuliani

Photographs copyright © by John Tertuliani
Photographs on pages 21, 239 copyright © by Ian Tertuliani

All rights reserved. No part of this book maybe reproduced, stored, or transmitted by any means without the prior written permission of Lotic Books.

First Edition

International Standard Book Number: 0-9761159-1-3

Library of Congress Cataloging-in-Publication Data

Tertuliani, John.
    Catching Bluegill / John Tertuliani.
        244 p.   22 cm.
        Includes index.

    ISBN 0-9761159-1-3

      1. Fishing.   2. Bluegill. I. Title.

                                              2006906495

Printed in the United States of America

For Patty, Tony, and Ian

CONTENTS

INTRODUCTION 3

1. LIFE HISTORY 7
2. TACKLE 23
3. VEGETATION 65
4. FINDING FISH 79
5. FLY TACKLE 105
6. FLY-FISHING 123
7. NEW ANGLERS 139
8. LAKE STRATIFICATION 159
9. LUNAR PERIODS 167
10. ICE FISHING 173
11. COOKING 197
12. QUICK TIPS 211

THE AUTHOR 239

INDEX 241

INTRODUCTION

Tall in stature with long fins, bluegill fight harder for their size than their cult-status relative, the largemouth bass. And taste? Many anglers do not know what they have been missing. You will be challenged to find a better tasting fish in freshwater than the bluegill. Yellow perch may be the only contender for such delightful table fare.

Too often their reputation with the public brings a vision of stunted fish begging in a small pond, which is desirable for kids just starting out. Fishing for bluegill is not as easy as you think. If you target the wary adults, you will be surprised how difficult catching them can be; it requires deliberate approaches. You cannot cast bait to water and expect instant bluegill, big ones no less.

Bluegill provide the opportunity to fly-fish that cannot be matched. You will often see a wake plowing water toward your dry fly. Wake forming strike or not, the take will not be a subtle slurp, rather a surface jarring strike is more like it. And when you fly-fish, you will not suffer the high brow of an amateur naturalist trying to tell you the shoe size of the nearest mayfly.

If you like live bait, they do too. If you do not like to use live bait, just throw a small plug or a jig; the results can be the same.

Bluegill are prolific, too productive for their own good; their success with spawning can easily fill small lakes and ponds. If a body of water becomes densely populated because bass are over harvested and bluegills are under harvested, the food supply fails to nourish all of the fish, this leads to stunted bluegill in the population.

You may have been too young to remember the first fish you

caught—you are not alone in this case—but it was probably a bluegill. If most anglers got hooked by catching bluegill, finding the willing fish within casting distance from shore is the reason.

Many anglers enjoy fishing for bluegill. The species lives throughout the country, far beyond the native range in mostly the Midwest. A common fondness for bluegill stems not from great experiences, but meaningful ones, such as catching fish at those critical times when it mattered most, the times when an angler was young and impressionable.

An angler often became hooked for life after a few successful trips of catching bluegill. A successful trip back then did not mean trophies of a lifetime, just a school of fish lined up at the bank with napkins tucked in their gills. And bluegill are the about the only fish you can count on to do just that.

Catching bluegill for adult anglers is not the same as it was when we were young, but many things from younger days just don't seem to be the same anymore. But few adults lose their desire for a basket of big bluegill.

Fishing for bluegill has become tradition for some anglers. When the weather begins to warm after the cold of winter, big bluegill are what spring rituals are all about.

Bluegill fight hard and they're plentiful. Better yet is how they taste. Their meat is firm and sweet. What more could you ask from a fish?

Unlike the crowded conditions plaguing so many fisheries today, you may have the place to yourself. One more thing, no one gives a damn if you enjoy eating them. You will not be stared at and lectured for taking a mess home to eat. I am glad bluegill are fried and not worshipped.

Hand-sized bluegill are game fighters and not as easy to catch as one might think.

1

LIFE HISTORY

The northern bluegill sunfish, *Lepomis macrochirus*, is a member of the family, Centrarchidae (freshwater sunfishes and black basses). Commonly called bluegill, it is also called bream or perch or sunfish, depending on where you live. The largemouth bass and the smallmouth bass are black basses.

Bluegill have distinct features. Most recognizable is their shape. A thin, rounded shape and colorful appearance makes the bluegill one of the most recognizable species in freshwater.

Adding surface area to their vertical form are a long dorsal fin and a pair of large ventral fins. As anyone who caught an adult bluegill knows, the fish pulls hard, often swimming in circles when it cannot get away.

Color tone varies. One body of water may make the colors brighter, while another may cause darker tones. Water quality (color and clarity) and the bottom structure (light-colored sand, dark rocks, off-colored plants) influence the color tones of a bluegill.

The prominent colors are an olive back with several dark bands running down the sides. But, the dark bands are not always pronounced, if present at all. Genetics and the environment (water conditions as described above) determine the patterns of color exhibited on a fish.

More distinct in appearance are the colors on the sides and breast. Bluegill are called dimorphic because the color of the male is different from that of the female.

The male can have colors varying from light purple to bronze coming down the sides to the breast. The female colors vary from light olive to bright yellow. The male breast varies from orange to copper. The female has a yellow breast, which is brighter and more yellow than the color on the sides.

Last but not least of the characters that define a bluegill are the flexible flap on the gill cover and the powder blue lower jaw. The gill flap is a uniform black in color. Other species of sunfish have a black flap as well, but the distant edge of the flap is a different color, such as red, orange, or shades of white. The longear sunfish has a gill flap of uniform color, but it is greatly elongated.

Bluegill behave much like largemouth bass when selecting habitat. The types of habitat that attract largemouth also attract bluegill. This should be no surprise; both species belong to the same family. Largemouth bass and bluegill are a classic example of predator and prey living in the same locations.

One nice thing about fishing for bluegill is you do not have to worry about where to look for them, just focus on vegetation. Bluegill love vegetation; it provides security from predators and a source of food living on and among the plants. What bluegill do not like is fast-flowing water in rivers and streams. They prefer the slack-water areas of rivers and streams.

Vegetation is not the only habitat that attracts bluegill. You may have caught them on rock piles while fishing for walleye. At times they suspend in open water, away from their preferred habitats.

Bluegill have many predators; common ones include the largemouth bass, northern pike, catfish, and wading birds such as the great blue heron. As a bluegill grows to a length of 6 inches or more, the chance of being eaten diminishes. The chance does not completely diminish. I have reeled in more than one 8- or 9-inch bluegill—only to have it struck by a huge bass—often at my feet while standing on shore.

As a general rule, smaller fish (less than 7 inches in length) stay closer to shore. Smaller fish may live clear out to the middle of a lake if shallow water and plants are available. Chances are these bluegill were born nearby.

Bluegill by nature live in groups. Schooling is a protective behavior exhibited by fish often preyed upon by other fish or predators.

Spawning is a perfect example of their social behavior. They spawn in colonies, covering an area with numerous nests built in close proximity to each other.

Another general rule: The larger the fish, the smaller the school. If for no other reason, there are fewer numbers of large bluegill. You may also notice the sizes of bluegill within a school are similar. You will catch the odd small or large one, but overall if you are catching bluegill smaller than you like, it's time to move and find the fish you are hoping to catch.

When you catch a single fish in deeper water away from shore, it is a sure sign that the oldest adults are large enough to be out on their own. Size may make the bluegill more solitary when not spawning.

When you catch several large adults from a single location, it shows that the lake or pond supports a healthy population of bluegill. A population is balanced by high numbers of adults. Spawning is more successful when large fish perform most of the spawning. The larger the fish that are spawning, the more successful is the transfer of efficient genes to succeeding generations. If high numbers of smaller fish begin spawning, the genes passed encourage a stunted population.

SPRING

Spring is the favorite time of year for many anglers living in the temperate latitudes. Having been stuck inside for months, an angler jumps at the chance to get out. Warm weather never arrives soon enough. Increasing periods of daylight causes the temperatures to rise. Warm air brings out the anglers and warm water brings out the bluegill as they move closer to shore to find warmer water and food.

Spring also happens to be the most predictable time for bluegill. A time when both males and females swim to shallow water to spawn. As with most species, the males move in first, claim territory, scrape out a nest, and wait for the females. The urge to spawn makes the prespawn and spawning periods excellent opportunities to find the

Spring is the favorite time of year for many anglers.

largest bluegill within easy reach.

The nests vary in size, depending on the behavior of the male building it, the nesting material, and the location. The more nests in the colony, the closer to each other they are built. Often the nests become smaller as the nests crowd together in a honeycomb fashion. The larger males either build their nest first or gravitate toward the center, by force, knowing their progeny are safer in the middle of the pack. The depths that the nests are built vary with the habitat. Most bluegill make due with what is available in water about 1 to 3 feet deep. The depth will vary with water clarity and the availability of desirable spawning materials at these depths.

Water temperature drives the spawn. Spawning begins when the water temperature reaches 66-68 degrees Fahrenheit. Warming water speeds the development of the eggs. Cold fronts delay spawning and egg development by causing the temperature to drop. When temperatures remain cold, the length of daylight (photoperiod) kicks in to stir spawning activities. Longer days mean spawning time is near.

Water levels can affect the success of spawning. High water increases the amount of spawning habitat available to bluegill. Low water can destroy the eggs by exposing them to the air.

Flooded shoreline often includes a diverse array of habitat. The increase in surface area provides more opportunity for the bluegill to spawn. Raised features such as vegetation and rocks and wood can provide physical barriers to help the male protect the nest.

The quality of the spawning habitat, high water or not, depends on what covers the bottom of the lake or pond. Clean sand is the preferred substrate for spawning. Small-sized gravel also provides an efficient bed if sand is not available.

The larger the material (such as sand grain versus gravel or small rock), the more difficult it is for the males to scrape a nest out of the bottom. Fine-grained material such as sand can be shaped and molded much easier than coarse material. Coarse material is hard on the adults. Scrapes and abrasions weaken the adults and expose their skin to ever-present strains of bacteria.

Eggs laid in a coarse nest can fall through the cracks. The eggs that fall through the cracks in the gravel and rocks do not benefit from parental care. A male fans the nest to keep the eggs free of silt. Eggs covered with silt do not receive enough dissolved oxygen to survive.

SUMMER

Summer is a stable period in the average lake and pond. Precipitation is not as prevalent as it is in the spring, but thunderstorms can produce heavy precipitation in local areas.

Water temperature in the summer is somewhat stable as well; the range in temperature does not change as much as it does during the spring. The temperature in the spring goes from freezing to warm water. During the summer the water temperature stays warm until the autumnal equinox.

Summer is a stable period for bluegill, a time when they rest after spawning and eat to restore energy.

Bluegill behavior during the summer depends on the body of water where they live. Within each body of water is a limited amount of food and shelter. Where exactly the bluegill go after spawning depends on where these resources are available. Bluegill will travel in order to find food and shelter. Recreational travel is not a viable option for bluegill to prosper; it is a waste of energy to travel farther than necessary. If anything, unnecessary travel poses risks to survival through a waste of energy and exposure to predators.

After laying their eggs, the females move back to deeper water. They spend the summer feeding, replenishing their resources used to produce eggs and spawn. They use available habitat in water deeper than where they spawned, following the food when necessary.

A feeding foray can last a matter of minutes each day. Feeding occurs at opportune times such as just after sunrise and before sunset. A feeding activity can last longer, such as a day or days of suspending over a large flat covered in sand—a place where midges hatch in the early summer.

After the male guards the nest and protects the sac fry, he stays in the shallows near the nesting colony, hoping for another female to show when the nest is empty. A male will spawn more than once if he is able to coax a gravid female into the nest. No surprises there.

The fry stay in the shallows as well, seeking protection from predators. As the fry grow larger, they venture out from the shallows, but tend to cling to habitat for protection, regardless of depth.

Vegetation is the preferred habitat of bluegill, but not their only choice. Rocks, woody debris, docks, and combinations of these features will attract bluegill. Plant beds in particular attract bluegill in the spring. Often called weeds, plant beds are important habitat because they provide bluegill with food and shelter.

The stalks, stems, and leaves shelter the young from predators, wave action, and exposure to direct sunlight. Rocks in shallow water and especially those close to shore, provide shelter as well. Young-of-the-year bluegill are easy prey when swimming out in the open or suspended over deep water. Plant beds will be discussed further in Chapter 3.

AUTUMN

Autumn is a less predictable time for bluegill fishing. Water temperature is the key. When the temperature drops, bluegill as a rule move to habitat in deeper water. Warm temperatures bring them back up to shallow water. Sharp drops in temperature can send them down until the temperature rises again.

Bluegill recognize the days becoming shorter. Decreasing periods of daylight bring decreasing air temperature. Water temperature decreases as well. They eat while they can to store reserves for the coming winter. Being cold-blooded, the decreasing temperature slows their metabolic rate, which affects their feeding activities.

As the temperature continues to drop, bluegill move toward their wintering areas. When water temperatures fall below 50 degrees Fahr-

Autumn is a time of change, a period of back and forth before winter.

enheit, feeding activities become shorter in length and less frequent.

Movements by bluegill through the seasons often reflect the size of the body of water. As mentioned earlier in the chapter, the distance and frequency of movement depends on the amount of habitat available and where it is located.

Autumnal movements may span a considerable distance if the lake is large and the habitat is scarce. The opposite may occur if habitat is available near deep water and food is plentiful. Bluegill often spend much of the year in or near confined areas such as a plant bed. But, not all the bluegill in a population will act in unison. There will be a small portion that does not follow the schools.

Autumn can be as productive as the spring if you know the lake bottom well enough to determine where the bluegill are most apt to spend time feeding or resting. Finding them then will become a matter of locating shelter close to a supply of food.

The familiar places for the bluegill to find food are changing. Their prey are also changing location with the changing weather. It is that invertebrates are migrating; rather they have completed their life cycle or are digging in for the cold temperatures to come. The bluegill may still feed in the same general location, but may spend inactive times located farther away from a food source. When conditions such as temperature, cloud cover, and wind are right, bluegill will come out of what is normally deeper water to feed.

WINTER

Winter is a slow time for bluegill, as it is for most species of freshwater fish. Bluegill are sluggish in cold water. Their metabolic rate is not as fast as it is in warmer water.

Bluegill will eat if little or no effort is required. Activities require energy to search for and capture their prey. Simply inhaling the food at their present location is the most efficient way for them to eat. This means you have to find the fish to put the bait in their face.

Bluegill do not remain still day after day, waiting for their next meal. Food for the bluegill is not abundant during the cold months of winter. To get around this they move around during the day as temperatures slowly change in the body of water where they live.

If you stay in one place for a long time, the bluegill may eventually find your bait and take advantage of a meal dangling in their faces. When a school passes by, the bites come one after another for a short period of time. Some bites last longer than others. The number of bites may depend on the size of the school.

Staying in one place for a long period of time is not the most effective way to fish through the ice, but there will be times when drilling holes until you find the fish is not practical or possible. Another reason for staying put is your confidence in the location. If you are comfortable where you are, you won't mind waiting as you enjoy your day on the ice. For more information on ice fishing, read Chapter 10.

If the biting lasts for an hour to an hour and a half, I check the

Winter can be a tough time to fish, but it can also be productive.

lunar periods for the day if I do not already know the times. The reason I do this is to see if the period coincided with the time the bluegill were feeding. For more information on lunar periods, read Chapter 9.

OTHER SUNFISH SPECIES

The redear sunfish gets its name from the red rim on its gill flap. Both the bluegill and redear live in the same general areas of a lake, preferring the same types of habitat; but the redear tends to take positions in slightly deeper water than the bluegill. Unlike the bluegill, they do not readily take artificial lures. Called shellcracker in the South, the redear eats snails, which may explain why it grows to a larger size than the bluegill.

Green sunfish are common in the same waters you find bluegill. Aggressive fish, they will strike harder and less reluctantly than bluegill.

Here is a bluegill x green sunfish hybrid, also common in bluegill habitats. Aggressive and willing, they pull HARD.

The pumpkinseed sunfish is more colorful, yet less common than the bluegill.

Rock bass are more common in rivers and streams, but will also be found in lakes and ponds. Similar to the green sunfish, they are aggressive and game for any bait offering.

The author holding a channel catfish that did not mind the cold rain of November or the bread on a circle hook. Fishing for bluegill is rarely limited to catching bluegill.

You just never know what you will catch in a bluegill pond. If you catch a snapping turtle, cut the line close to the mouth with a fast swipe of a knife. Pull the line tight before making the cut from behind the turtle.

All you need to catch bluegill is an assortment of floats, split shot and sinkers, and hooks. Simple, yes, but float fishing with live bait is one of the most effective ways to fish.

2

TACKLE

A cane pole will do. But rare is the cane pole today. Times have changed, so has the tackle. Bluegill are still the same, unless the pressure gets to them. Public fishing pressure can condition bluegill to look long and hard, then refuse to take the bait. As pressure increases, their ability to avoid a hook increases. Tackle needs change with changing conditions and bodies of water.

Finding the fish is the most difficult step. The bigger the bluegill, the harder they will be to find. Larger fish occur fewer in number. Having fewer adults in the population is normal.

Once you find them, then, you have to present bait that they will bite without hesitation. Triggering a strike without hesitation is not difficult, even from large adults. You do not have to match the hatch to get their attention. Presenting bait will earn their undivided attention as soon as they see it. Whether they strike or not depends on how well you present the bait. Bluegill will strike the offering if they recognize it as food.

GETTING STARTED

Tackle, like the rod and reel, can be as simple or as high tech as your desires dictate. If want to keep things simple, live bait and any outfit will do. But, if you love tackle, as most of us do, you can use slip floats, plugs, jigs and spinner baits and the rod and reel of your choice. Tackle should be what you already own. There is no need to buy special tackle just for bluegill, unless you are serious for bluegill and panfish and want the most effective tackle possible. The most effective does not mean the most expensive. Remember, cane poles used to get the job done.

Tackle should be chosen to suit the needs of your fishing. Are you fishing from a boat? From shore? Do you fish in ponds? Do you fish in public water that receives relentless pressure?

Light tackle is recommended for bluegill. With such a small mouth, they need tiny baits to attract their attention. Their large eyes provide excellent vision to feed on minute prey, vision that can recognize heavy tackle. As a rule, the more popular the water, the more difficult it will be to catch big bluegill on a regular basis. Light to ultra light tackle will present tiny baits so realistic that even conditioned adults will be fooled.

POLES

Still popular in the South, poles are the easiest way to fish live bait. But, like all other tackle, technology has improved the simple concept. Graphite, fiberglass, and composite rods have replaced the cane pole. Varied in length, they are one piece and telescopic, making a modern pole ideal for plunking the bait straight down through the heaviest vegetation.

Poles are popular with anglers who fish from boats in areas where casting is not necessary. Fiberglass poles are popular with crappie anglers who fish among fallen trees. Graphite poles are popular in Europe where much of the fishing is done from shore. European poles have several sections to reach out from shore.

When fishing with a pole, the distance is restricted to about the length of the pole. The length of fishable line is as long as the rod. This length is the distance from the tip of the pole to the butt section. If you make the line longer than the butt, the fish will not come back to your hand when you swing the line back.

If you are stringing a cane pole with two sections, tie the line to the bottom of the tip section just above the connecting ferrule. Attaching a line at the tip of a cane pole increases tension on the pole at one point and increases the possibility that the tip will snap off.

Once popular, cane poles are rare today.

If carefree fishing is your pleasure, spincasting equipment is for you.

SPINCASTING

Tackle that is easy to use is important when starting out. The most simple to use is a pole with fixed line, next is spincasting tackle. Many anglers fishing today started with spincasting tackle. Spincasting reels are as carefree as can be. No worries about backlashes and tangles. But, when not used properly, spincasting tackle can tangle just as easily as any other tackle.

Spincasting rods and reels come in various levels of performance, from practical combinations for beginners to high-quality outfits for serious anglers who want the best tackle available.

Choose a spincasting rod with a length from 5'6" to 6'6". A casting rod can be used with spincasting reels and baitcasting reels. Because of this, choices in casting rods are endless. Look for a rod marked light to medium in power.

Construction materials vary from fiberglass and graphite composites to high modulus graphite. High modulus graphite, called "pure carbon" by the Europeans, is super sensitive. The high cost should be no surprise.

Spincasting reels have come a long way in quality and performance. Bearings come standard in most spincasting reels and stainless steel components are put in the more expensive models. Two noticeable differences between a spincast and spinning reel are the nose cone and the push button release. Spinning reels have an open face with exposed spool and a semicircular bail to cock instead of pushing the button and holding it down with your thumb.

Look for the option of left-hand retrieve available on some models of spincasting reels. If you right-handed and do not like switching hands after making a cast, a reel with left-hand retrieve is for you. A left-hand retrieve eliminates the need to switch hands with every cast.

SPINNING

Spinning tackle is the most popular equipment for bluegill and numerous other freshwater species. One exception is the tackle used for largemouth bass. Baitcasting tackle is more popular with bass anglers and heavy tackle at that.

Numerous spinning rods are sold for crappie and panfish, but not marketed for bluegill alone. The line weights appropriate for a rod are marked on the blank just above the handle, along with the power. Choose a spinning rod with power ranging from ultra light to medium. Action is also marked on a high-end rod. Action ranges from slow to fast.

A rod with a slower action is more suited to light lines because more of the rod bends. Instead of worrying about finding a rod with a slower action, just look for a longer rod made from composite materials. Composite rods tend to have actions slower than graphite rods made from high modulus blanks. A slower action (softer tip) is needed when using live bait. You are less likely to rip the bait off when casting and setting the hook with a slower rod.

The fixed-spool reel makes spinning tackle a long-distance casting machine. Casts are made with the index finger controlling the line.

Live bait rods normally range from 7' to 9' in length. European rods used for match fishing are designed for bait fishing and available in extreme lengths. A European rod 12' in length is not necessary to be effective with live bait. This length would prove difficult to use in the confined spaces of a small boat or canoe and when fishing with other people. A 9' rod is a challenge.

Spinning reels offer numerous features, options that make fishing a pure pleasure. Multiple bearings machined from stainless steel, oversized drags, and technical spools mean you have choices. Spinning reels come standard with left-hand retrieve.

BAITCASTING

Baitcasting tackle has undergone incredible changes in performance and quality. Most changes have come with the evolution of professional bass fishing. Anglers demand the best as pros in any sport do. To serve these markets, tackle companies release new models every year. The quality and performance can go way beyond the needs of anglers who do not compete in tournaments.

Baitcasting reels have also been called level winds. Look for models designed for light lines. A few companies make reels with spools designed for 8-pound test line. If you look in a fishing tackle catalog, you will see 10-pound test as the lightest line weight for a specified capacity. You want to know the line capacity for light line so you don't buy a reel with a high-capacity spool.

Avoid a reel with a high-capacity spool; it will have a thin arbor. You will use twice as much line to fill the spool to the rim. And, monofilament line coils more on a spool with a thin arbor. You can use lines as light as 6-pound test line on an Abu Garcia Ambassadeur with good results.

Baitcaiting tackle is popular with bass anglers, but heavy for bluegill. If long casts are not required, such as when fishing from a dock or boat, then baitcasting tackle will work just fine.

Baitcasting rods, like the reels, are designed primarily for fish larger than bluegill. As mentioned in the section on spincasting tackle, the selection of baitcasting rods is enormous. Medium-light action rods in 5'6" to 6'6" lengths will do nicely for bluegill.

Baitcasting rods come in two styles, casting and trigger. Casting rods have a pistol grip, the handle becoming broader toward the bottom. Trigger rods are uniform in thickness in the handle and are longer in length than a comparable rod with a casting handle.

A baitcasting outfit is not common tackle for bluegill, and it is not recommended for newer anglers. Baitcasting tackle is no more complex than spinning tackle, but it proves a challenge to use with light line and tiny lures. The spool unwinds to let out line. Spool tension is adjusted to the weight of the lure or bait being cast. Other reels described have fixed spools.

MONOFILAMENT LINE

Use a premium monofilament such as Trilene XL or Stren. Super lines and otherwise braided lines can be used, but use a monofilament leader to avoid spooking the fish. Under most conditions you will encounter, 4-, 6-, and 8-pound test monofilament will pull you through. You can pull a canoe back against a gentle current with 6-pound Trilene.

Trilene is extra limp, a line with low memory. A low-memory line does not form coils unless it is left in direct sunlight or heat for extended periods, which are two conditions that will get the best of even premium monofilament. Stren is a bit stiffer and a good choice if there are rocks or fallen trees in the area, where abrasion would be a threat to light line.

Experienced anglers with their reel drag properly adjusted will benefit from using 4-pound test. Use 6-pound test in heavy cover, such as plant beds thick with vegetation. For the toughest conditions use 8-pound test. Experienced anglers should not need 8-pound test very often.

An important habit to keep when using light or ultra light line is to retie often. Cut lengths of line from the running line if the line feels rough when you slide your fingers down it. Also, retie the knots after pulling hard against a snag or a fish temporarily hung.

If using light line concerns you or you hate to retie hooks and other terminal tackle, a heavier line such as 8-pound monofilament is warranted. Trilene XT holds knots well, and comes in low-visibility green. Clear line is recommended for most fishing conditions.

Spinning reels are ideal for fishing with light line. Proper filling with line is required to keep the reels functioning smoothly. Too much line is worse than not enough. The reel spool on the left has been loaded with too much line. Look at the mounds formed in the line near the top and bottom of the spool on the left. The spool on the right is properly filled, which is evident from the straight-sided layering of monofilament. A reel with too much line on the spool creates loops, especially light line, because the light tackle on the end of the line does not maintain sufficient tension to prevent slack. Excess line with slack in it leads to tangles, often requiring the excess line to be removed. Please pick up the line you cut from your reel and recycle or dispose of responsibly.

FLOATS (BOBBERS)

A float is not necessary but is recommended for live bait. A float will improve fishing for several reasons. First, the bait will be presented at the same depth with each cast, no additional effort required. The float does the work, so to speak. Second, the bait is suspended above the bottom where the bluegill are more apt to find it. Third, a float becomes the bite signal; a visible signal to pay attention to, and an enjoyable one at that.

More often called bobbers, floats have gained popularity through the influence of European match fishing. Fishing pressure on public waters in Europe and other foreign countries is so great that our methods would not produce as well as the European methods of fishing with live bait. Methods have to be so specific to be successful that finesse fishing doesn't come close to describing it.

Floats are perfect indicators of a strike. Who doesn't love to see one go down, even if it isn't yours? And a float presents the bait at the same depth each time. Once you find the depth at which the fish are feeding, dropping the bait to the same level with each cast is guaranteed.

Red and white floats, the round ones with the spring tops, are popular with many anglers. They also come in yellow and orange. Time was when choices were limited, but no longer. Floats come in many shapes, sizes, and colors, but a fancy float is not necessary to catch bluegill, lots of them. Use simple tactics and sparse tackle and you will catch bluegill—lots of them.

In-line floats are recommended over floats that attach at one end. The running line passes through an in-line float, making one more trouble free than the waggler style that attaches at the tip of the float stem. Floats attached at one end tangle frequently when cast.

Floats lacking sufficient weight will not go under water with every strike. This is not productive; it gives mixed signals to the angler. Consistent results are necessary to learn the skill and enjoy the experience. And, improperly weighted floats are the surest way to have hooks swallowed without your knowing it until it is too late.

Floats are simple but important devices that increase your chances for success. No matter what style of float you use, weight it down until it can barely float. The goal is neutral buoyancy. A neutrally buoyant float is weighted until it is on the verge of sinking, when the slightest tension will pull it under. Bluegill do not feel resistance from a neutrally buoyant float. They will feel the resistance of an underweighted float. Large bluegill will drop bait with resistance if they receive considerable amounts of angling pressure.

The float moves as soon as contact is made with the bait. The float does not always go down; it may rise up slightly if the fish has picked up the bait and is eating it in place. Bluegill swim away with bait most often, but this is a behavior from feeding in a school of fish trying to take the food away from the captor. Either the float goes down or rises slightly; any motion at all is the signal to set the hook.

A float is not necessary but is recommended for live bait. Floats serve the angler more than to signal a strike. A float makes a precise presentation, time after time, removing the guesswork involved with returning to the same depth as the fish. Floats suspend the bait up in the water column where the bait is more visible and easier to smell.

Slip floats can be used at any depth, an advantage when fishing depths greater than 3 feet. As the length of line hanging down below the rod tip increases, the difficulty in making a cast increases, as well as the potential for a tangled mess. You may not discover this until you finally reel the line in, wondering why you have not gotten any bites in a long time. You retrieve the line to find the hook caught on the float, out of reach of the fish.

If using a fixed float, a cast with 5 or 6 feet of line out can be done, but you have to swing the rod as it is held high. You can then reel the line in up to the fixed float. Too much line may hang the hook just out of reach, which may become an issue when you have a fish on the line.

Another advantage is line control. With the bait suspended, the slowest, most subtle movement is possible. This is a great advantage for catching inactive fish, and fish in cold water with slowed metabolisms. A subtle movement may be all the trophy bluegill needs until it has had enough and strikes.

Floats are also an effective way to present plastic baits and jigs. Plastic baits and jigs are no different in their purpose than live bait; they are just fakes. But, few people fish plastics bait and jigs as live bait. There are no hard and fast rules in fishing, other than to give the fish what they want. What they want is something to eat, no more.

LURES (PLUGS)

Most any lure will work for bluegill. The most important aspect of your lure selection is size; it has to be small, not to draw strikes but to fit in the mouth of the bluegill. Big bluegill will strike bass lures, but you will not hook many of them.

A lure measuring 2 inches in length is about as small as they come. With this in mind, floating minnows are perfect for retrieving over and along submerged plant beds. Because the minnow runs shallow, it will swim above the plants. Bluegill love the sight of a lost minnow cruising over a plant bed; it's a target too easy to let swim by. You will like a shallow runner for its tendency to stay above the plants instead of diving into a snag. The edges of cattails are ideal locations to cast shallow running plugs in the spring and summer when bluegill are spawning.

The Rapala CountDown CD03 is a miniature minnow for fishing deep. This plug sinks well, as designed, and should be used in more open water as a search lure; it is prone to catching everything in the area, especially the largemouth bass. You may not catch many bluegill for the bass.

The floating minnow casts well with 4-pound test line. The countdown version is heavier minnow and can be cast with 6- and 8-pound test. Spincasting and spinning tackle are more efficient casting such light lures.

Many companies make tiny lures about 1½ inches in length; most styles are fat bodied. The fat-bodied design creates more pressure waves by displacing more water than a minnow bait with thin form. Fat bodies are more difficult to eat, but this should not discourage you from trying some, many anglers have great success with them.

If you are using a floating or shallow running plug, vary the retrieve. Let the plug float back to the surface and sit for a minute, then twitch, twitch, and stop. Twitching is often needed to coax strikes. If you can catch fish by simply reeling in, do it.

Sinking and deep diving plugs are reeled in much of the time, but can be jerked to mix up the presentation. Active bluegill are out feeding and need little encouragement to strike. Seeing the bait is often enough to trigger a strike, but it is more accurate to say artificial lures need movement to attract strikes because you are relying on sight and sound to draw attention.

SPINNER BAITS

The spinner bait is such an effective design because the spinning blade creates a sparkling visual image and produces sound. Using a scented grub on the jig adds smell for a third attraction. Spinner baits come in two forms. The safety pin style is represented by such lures as the Johnson Beetle Spin and the horse-head style such as the Blakemore Road Runner. The Road Runner is more popular with bass and crappie anglers, but works well with bluegill.

Spinner baits can be fished shallow and deep, fast and slow. They offer more options for presentations. These single-hook lures do not snag nearly as much as lures equipped with treble hooks. One thing about spinner baits is they are not carefree baits. They twist the light line and need a swivel snap to reduce line twist. And every so often the line wraps the spinner arm of the safety pin style. This is not a great problem, but a nuisance at times.

The 1/32-ounce spinner bait in either style is light; even with slow speed it will stay at the surface. But the 1/32-ounce spinner bait is too light for many fishing situations. Clear water, shallow and dead still is a situation in which the 1/32-ounce spinner bait excels. Ultra light tackle will be needed.

The 1/16-ounce spinner bait is an all-around weight sufficient for most water you will encounter while fishing for bluegill. Light tackle will serve your needs for a rod and reel spooled with 4-pound test line. The tendency is to reel spinner baits in too fast. Your presentation should run through the lower one half to two thirds of the water column. Slowing down your reeling cadence will allow the spinner bait to run deeper.

The 1/8-ounce spinner bait is heavy, but may be necessary in deep water or waves. This weight tends to dredge the bottom of most bluegill waters, but will be necessary in some lakes. Dredging the bottom is not a bad presentation, but keep in mind that bluegill are not bottom dwellers. The most active fish will be up in the water column looking for food. When the cover is deep, such as rock piles and points, the 1/8-ounce spinner bait will do well if the hook is small.

JIGS

If one lure can be considered more popular than spinner baits, it is a jig. First used in saltwater, jigs offer many options to catch fish. A jig can be dressed with hair, feathers, synthetic fibers, plastic, scented plastic, and live bait. Combinations of these dressings are used when fishing conditions are challenging.

Jigs sink faster than spinner baits in the equivalent weight. The lack of a spinning blade and wire arm makes a jig more streamlined. Because of this, jigs do not ride as high in the water at the same speed. Fishing close to the surface is not as easy as it is with spinner baits. Jigs are not as snag resistant as spinner baits. The differences between spinner baits and jigs are not disadvantages, although there is an obvious difference in the performance of the two lures. Both are extremely effective. You should try spinner baits and jigs; you will like both, but you will more than likely develop a preference for one over the other.

When bluegill are actively feeding you can catch more fish with lures instead of live bait. Lures make it possible to reach more fish. Live bait requires time spent baiting hooks, then waiting for a bite. With a lure, the angler makes a cast, then another and another.

HOOKS

Hooks are important. Fishing with live bait is simple, yes, but the terminal tackle you use should be high quality. A hook is the last place where you want to save money. Fishing starts with the hook. The hook is the direct contact you have with every fish you catch. Think of trying to catch a fish without a hook. Then, think of using a cheap hook on a trip you have waited months for.

Hooks come in more shapes and sizes than you will ever need for bluegill, but three basic shapes will be featured here. If more variety is desired, other options in hooks are available.

The first shape, a popular style for live bait, is the long-shanked, thin-wired Aberdeen design. A modification of this design is the Tru-Turn hook by the TTI-Blakemore Group. Invented in 1959, the shank is offset to turn toward the pressure. The Tru-Turn works particularly well for bluegill.

The second shape, an opposite design of the long and thin, is the Ultra Point Bait hook by O. Mustad & Son, Inc. Mustad has been making hooks for over 100 years, needless to say they do it well. The Mustad Ultra Point represents high quality in the purest sense. The Neon Bait hook is perfect for small baits and bluegill.

The third shape is the circle hook, a recent design for freshwater sport fishing, which is becoming more popular for live bait fishing. Catch and release is not successful if the fish swallow a hook. Responsible anglers realize this and have begun using circle hooks for saltwater species and the black basses in freshwater. For the Canadian provinces requiring barbless hooks, the governments would do their fisheries a greater service by requiring circle hooks. Barbless hooks are still being swallowed. Pierced guts and ruptured gills kill the fish. Barbless hooks make removal easier, perhaps after the damage is done.

The circle hook penetrates flesh after swinging around in the mouth of the fish. The hook eye has to change the direction from which it entered the mouth in order to expose the rounded point to flesh. This happens even if it is in the gut of a fish; tension on the line will put it straight out. Once the eye of the hook exits the mouth, the eye rotates toward the angler (the tension on the line), and the point is exposed to the fish; it is then that the fish is hooked.

The effectiveness of the circle hook only catching the jaw is remarkable. A fish does not swallow the hook because the hook cannot rotate until it is turning around the corner of the mouth. If the fish never swallows the hook, it still has to turn about 180 degrees to make contact.

Circle hooks will become more widely used in the future, but for now standard hooks are more popular because the hooking rate is higher, and freshwater circle hooks are almost nonexistent in sizes small enough for bluegill. Circle hooks may never achieve the hooking rate that straight points do, but decreasing the mortality rate is worth the number of strikes missed. What better excuse do you need to keep fishing?

One of the circle hooks tested for *Biotactics* is the Mustad Ultra Point Demon Circle, Extra-Fine Wire. This design is flat (the point is not offset), which is very important to the function. A flat design makes the hook virtually swallow proof. Another company sells circle hooks in small sizes, but the points are offset and the hooks are swallowed. The Demon Circle is the only design recommended for bluegill fishing, with both live bait and soft plastics. Size 6 is the smallest size available, just small enough for bluegill.

Sure you will miss some strikes when you switch to circle hooks. Most strikes will be missed not because of the hook design, but because of the actions of the angler. Circle hooks have to be allowed to work; most of us do not posses the patience to let the fish hook itself.

Letting the fish run with the bait is the hardest thing for an experienced angler to do. We have spent our entire fishing careers trying to set the hook as soon as possible. With the slightest hint of a strike we reach for the sky. This is counterproductive with circle hooks.

LIVE BAIT

Wax worms make ideal bait for bluegill. Easy to use and widely available at bait shops and well-stocked pet stores, wax worms are the perfect size for bluegill. Wax worms stay on a hook and they may have a more attractive scent from being fed beeswax.

Night crawlers are the most popular bait for many anglers, but for bluegill they are large bait, much of it wasted because the fish cannot swallow the bait; they bite off what they can and leave. Chunks of night crawlers will not attract wise fish. Fish in public waters receiving constant pressure learn to recognize a chunk of night crawler because they see it so often. In less-pressured water, night crawlers hooked once make attractive bait and catch who knows how many bluegill each spring and summer. Chunks work, too.

Garden worms, red worms, dew worms, and whatever else they are called, look like baby night crawlers. These worms are an ideal bait for bluegill, just like wax worms, they are a prefect size. You cannot find a garden worm too small for bluegill. Bluegill love them.

Crickets are very popular in the South. Crickets make outstanding bait for bluegill, no doubt, but they are one-bite bait, expensive, and not as available as other baits. Crickets are so delicate you may cast them off. Using them with a pole as often done in the South is not a problem.

Natural baits are effective, but you may have to collect them yourself. Any aquatic larvae you can find will work well. Grubs dug in your yard or found in the forest will get undivided attention. Beetles and grasshoppers work well if you can gather enough to use. Aquatic insects and invertebrates such as crane fly larvae, hellgrammites, and leeches work extremely well.

Food is another option. Lunchmeat, pieces of hot dog, and bread will work in a pinch. Use fresh bread, knead a small piece the size of a wax worm and hook it once as you would live bait. White bread is preferable because the fluffy texture is easy to knead into a small roll. Whole-grain breads do not use refined flours and may not stay on a hook as well as white bread.

Floats come in many shapes, sizes, and colors. The most familiar is the round-plastic style. A brass wire passing through the float clips to the line at both ends. Push the button down to clip the line with the bottom clip. Then, put your finger over the bottom clip and line and push the button down again. This will force the wire up through the button so the line can be clipped with the upper clip.

Add enough weight to sink the float to the upper color, so the lower half is under the water. A float weighted more than half under water is even more sensitive. As long as some portion of the flat stays above water, you will be fine. The goal is to submerge as much of the float as possible, without sinking it. Floats will take more weight than most people use.

A simple but effective rig for shallow water presentations is a float, weight, and hook.

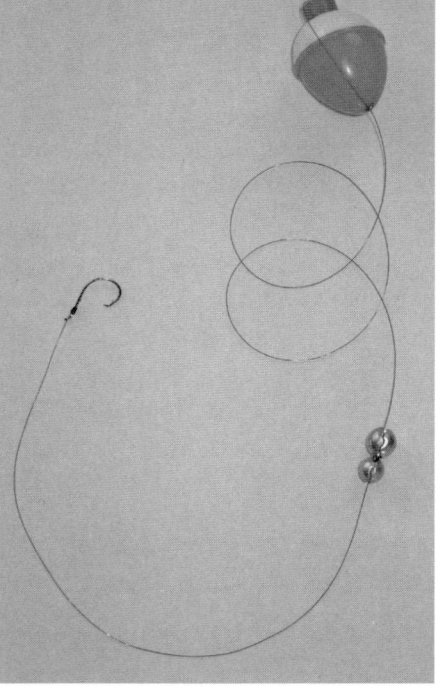

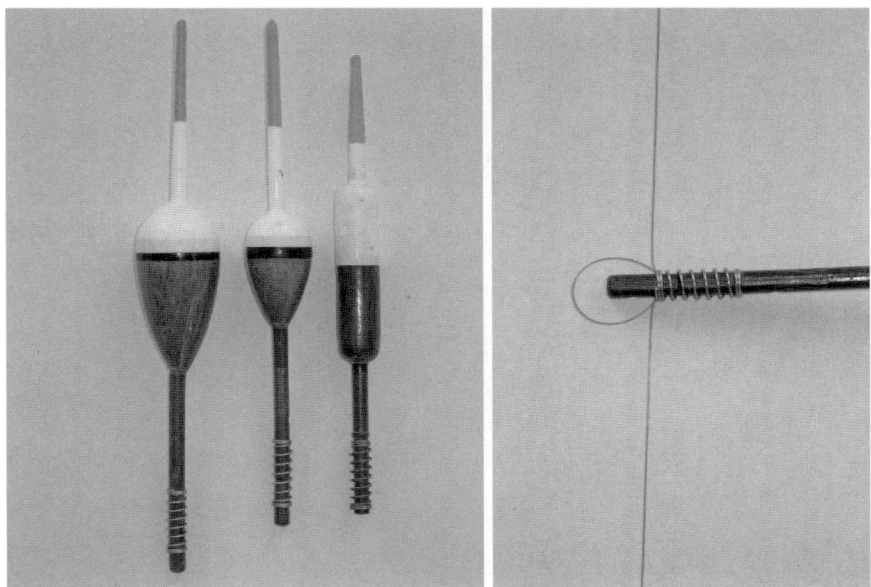

Spring floats are made of wood or foam and a plastic stem. Easy to use, these floats are ideal for shallow water (less than 3 feet). Attach the float; cinch the line one wrap around the float stem as shown in the upper right photo. How much weight is needed varies with the float. A float made of foam does not take the same amount of weight as one made of wood. Lead shot is denser than other types of shot. The number of split shot needed depends on the material of the shot as well as the float.

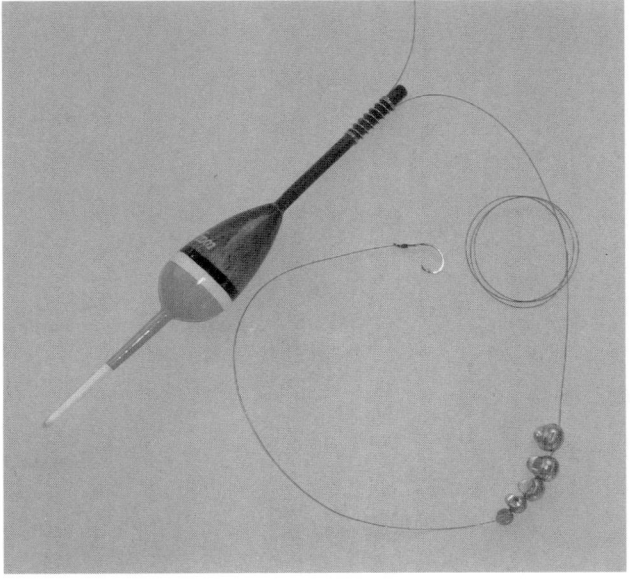

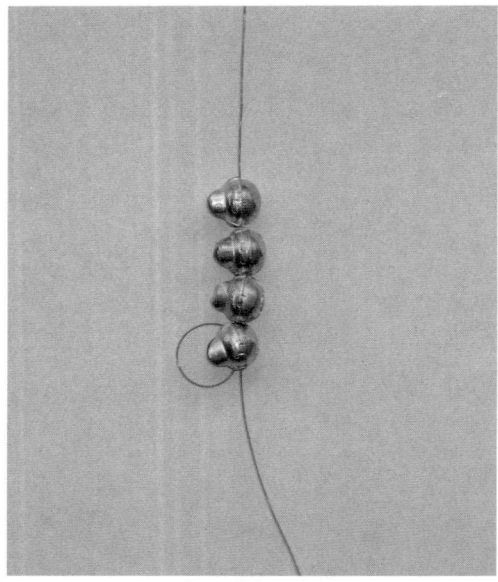

Split shot can be added and removed as needed, an advantage when using floats. Making a float neutrally buoyant is easier when you use small amounts of weight. Larger sinkers may not be the correct weight and may need to be replaced. There are a couple of drawbacks though. Split shot pop off under tension, and often slide down to the hook when cast. To prevent sliding, cinch the bottom shot with a wrap of line. Cinching does not keep the shot from popping off.

Slip floats slide up and down the line. The line passes through a pinhole at the top of the float and exits through the open tube at the bottom. The bulb of the float is often made of wood or foam. A slip float needs a stop at the top to set the depth. A stop (called either bobber or float stop) comes in three forms: a pre-tied knot on an installation tube, called a stop knot; an oval-shaped rubber bead, stored on a wire loop used for installation; and a tiny plastic strip with holes in it to weave the line around and bind it in place. A small plastic bead comes with each stop. The bead prevents the float from sliding over the knot. See also page 46.

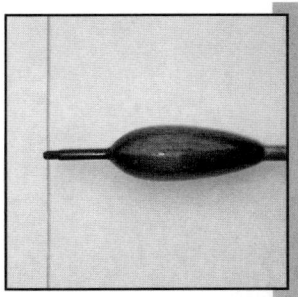
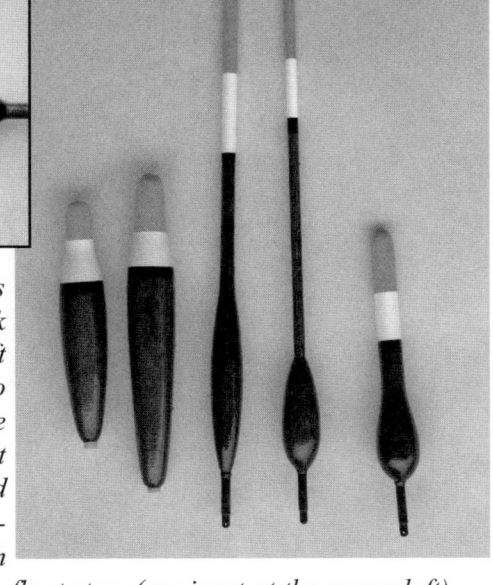

Here are the European versions of slip floats introduced by Mick Thill. The two floats on the left are called center sliders; the two long ones in the middle are wagglers; and the on the far right is the stealth. The waggler and stealth designs attach at the bottom only. The line passes through the bottom perpendicular to the float stem (see inset at the upper left).

All slip floats have the potential to tangle when cast. Split shot placement below the float can induce or reduce tangles. Experiment with the spacing between shot. Mick Thill has specific ideas for shot placement. Each float has the shot pattern marked on the back of the package. The weight distribution is divided into two groups of shot. The bulk of the weight is close to the float and then one or two shot are down near the bait.

I modify this pattern to use a slip sinker, then a swivel, then about a 10-inch length of line (used as a leader or snell) to the hook. The benefits of using a slip sinker are twofold: split shot fall off when tension is put on the line, which will happen when you pull the rig from plant beds and reel in a fish embedded in the plants, and second, tangles are less common with one weight. The swivel below the float is a benefit as it is a lower stop for the float and a tie link to attach the leader.

The secret is to find a slip sinker of correct weight for the float. A split is often added to make up the difference. I add a split shot above the swivel and not close to the hook. You can shorten the hook leader to 6 inches to put weight closer to the bait. The swivel can serve as weight instead of placing split shot close to the bait, thus defeating the modification.

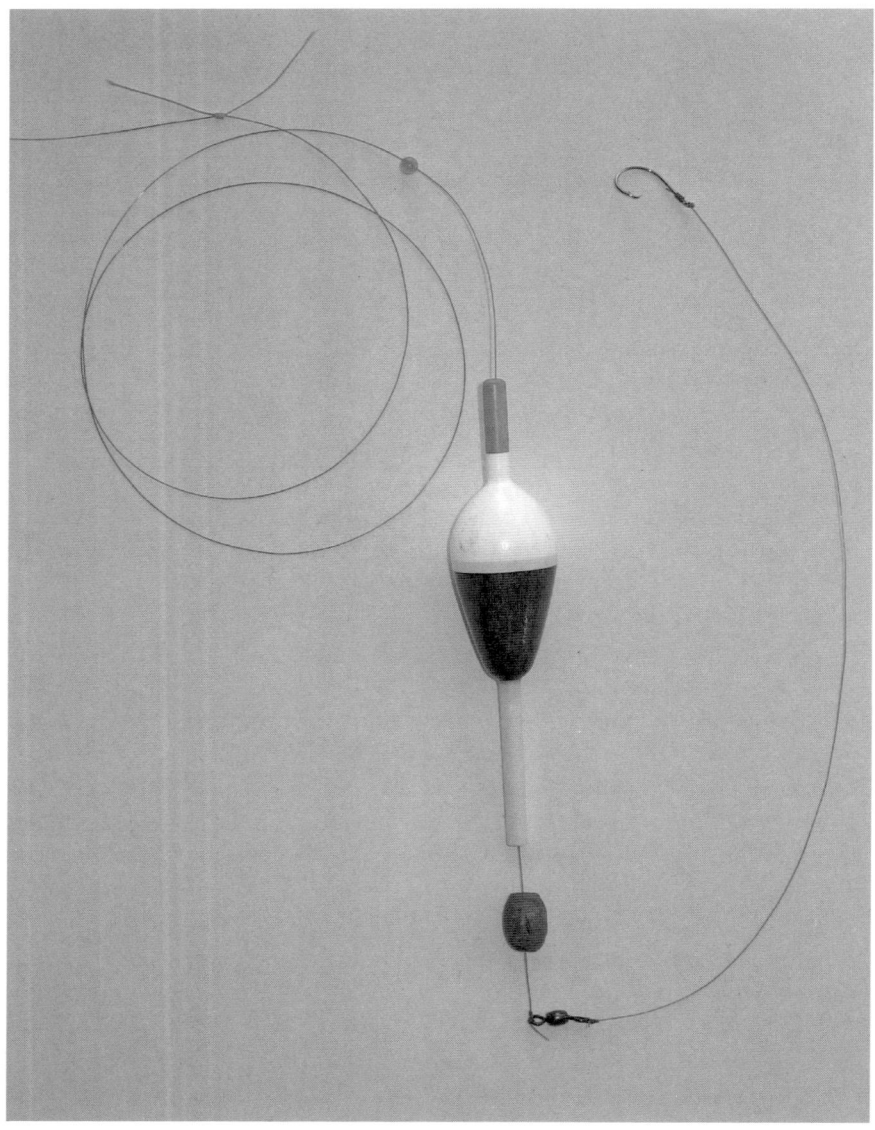

Assembling a slip float is easy. Most important is the stop. The stop is attached to the line first. If you tie your own stops, you can attach the homemade stop last. Slip floats are versatile; use one at any depth, even shallow water. The stop can be an inch above the float and work just as well. If you ever need to fish deeper, pull the stop up and go.

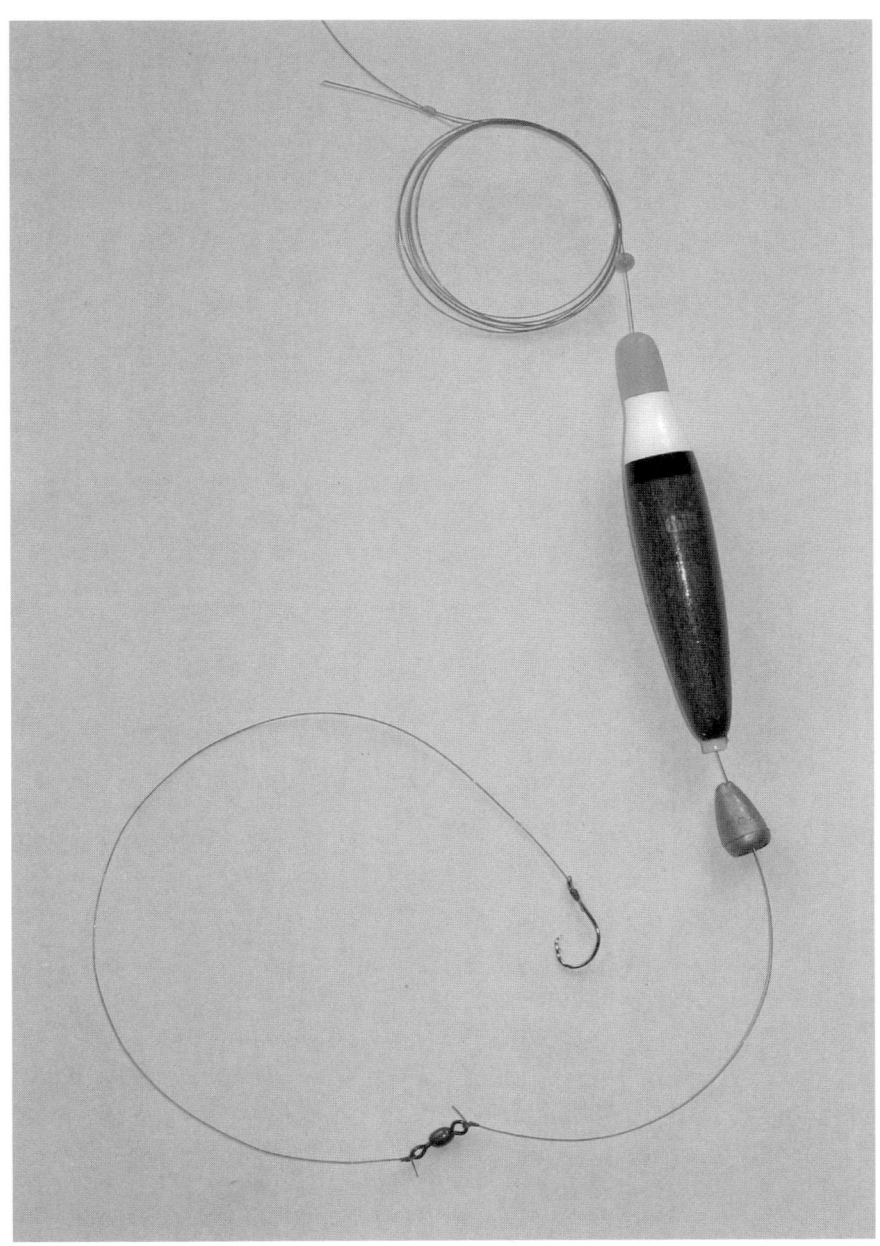

An in-line slip float is versatile and less prone to tangles. You do not have to mess with split shot placement; a slip sinker and swivel will do.

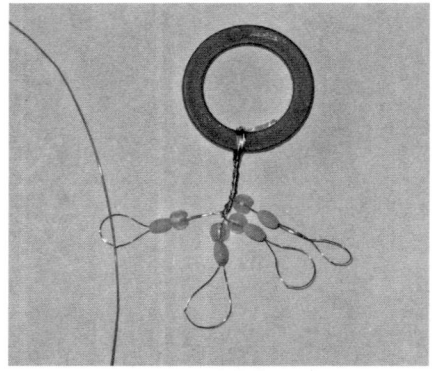

Float stops (bobber stops) come in several designs. The design on the left is made with pliable material resembling rubber. These stops are packaged on wire loops along with plastic beads. Put your fishing line through a loop of wire, and then slide the stop and bead over the line. You will have to pull the short end of the line through the stop. Then slide the float on the line.

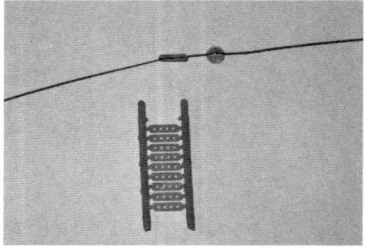

Another stop is a tiny piece of plastic with four holes in it. The line is weaved in and out of the holes while attached to the ladder. A solid stop, but unlike the rubber stop above, the position is fixed once in place.

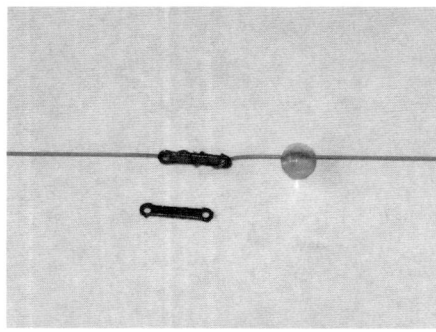

A Speed Stop is a sliver of plastic with a hole at each end. It's another solid, but more or less fixed stop. The line goes in the hole at end, wraps the stop, and then exits through the hole on the other end. (Photo enlarged)

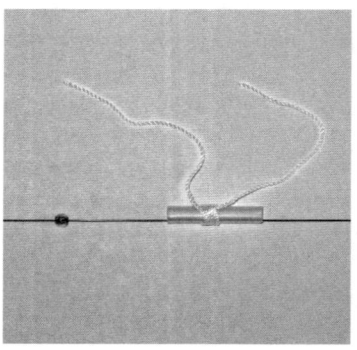

You can slide a string stop up and down the line to change the depth; this is a desirable option. Do not over tighten the stop knot; it will twist and curl the line when you slide the stop. Heat from friction may weaken the line. Stop comes loaded on a tube. Put the line through the tube, slide the stop off the tube, remove the tube from line, pull both ends and trim the excess. The bead goes on the line below the stop.

Float stops can be made with an overhand knot of several wraps. The string does not have to be a special formula for stops; you can use any braided line with success. Heavy thread used for carpet work will do, as will monofilament in an emergency. Monofilament has to be thick enough to stop the bead or the float. A bead is not absolutely necessary. If you look closely at some floats, you will notice that the holes in the floats are smaller than the holes in most beads that come with a stop. Newer floats have large grommets to prevent wear and these floats need a bead.

Making a float (bobber) stop

Step 1: Start with about 10 inches of braided line or heavy thread
Step 2: Wrap one end through the loop 5 times (wrap around line)
Step 3: Pull both ends snug (tighten when set at desired depth)
Step 4: Trim ends, leaving enough to tighten later with your fingers

Slide a bead on the line below the finished stop, then the float, add enough weight to make the float stand up and nearly sink, then the hook.

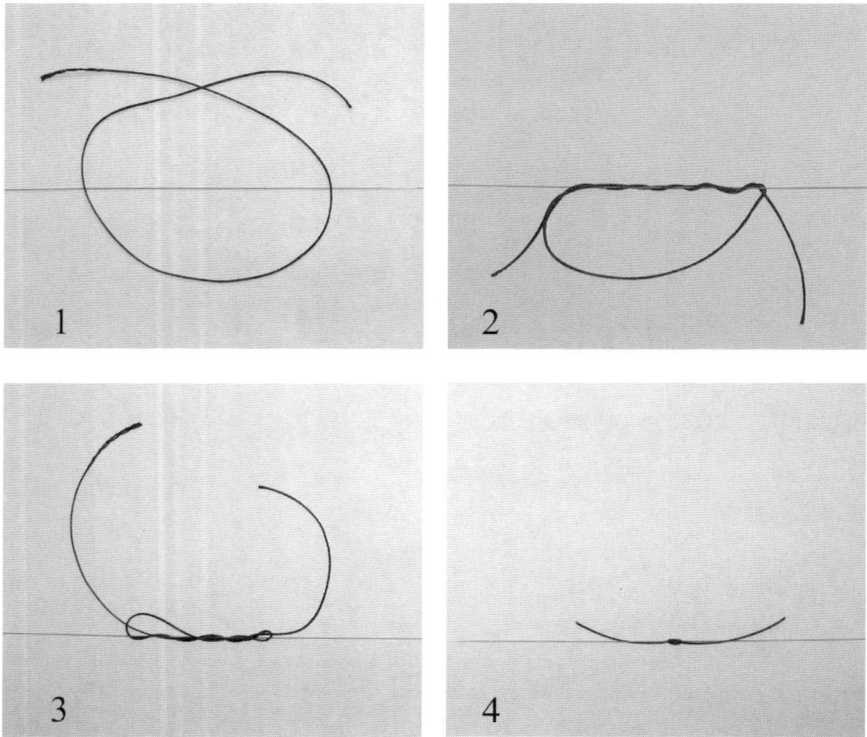

Making your own float stops means you can tie one on at any time and at any place on the line. If you use commercial stops, you will have to cut the line and remove all the terminal tackle so the stop can go on before the tackle. Any thread can be used to make a stop if thick enough. Monofilament can be used too, but monofilament should be used as a last resort.

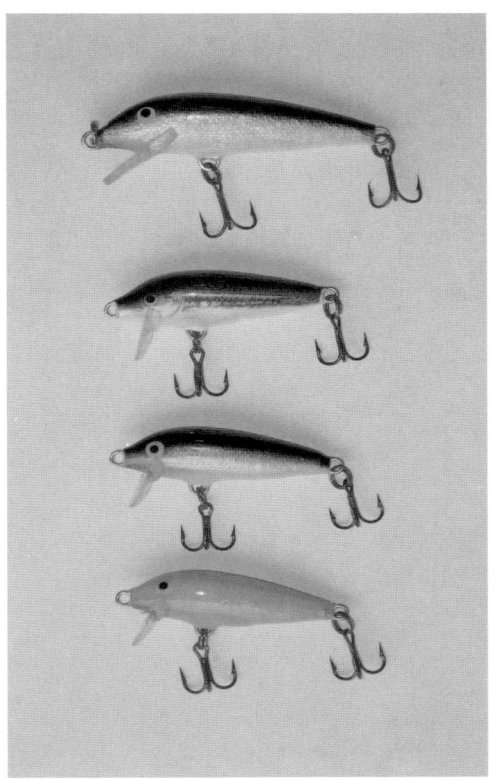

Minnow baits are ideal lures for large bluegill. Minnows are common in the diet of adult bluegill. Choose a minnow that is about 2 inches in length. Use floating minnows for over the top presentations on plant beds. Sinking and suspending minnows are effective along deep edges of plant beds, around wood, over rocks, and along steep banks—the places bluegill may be feeding or resting. Floating minnows are extremely light; 4- to 6-pound test monofilament and a 6', fast-action rod are in order. As fun and effective as the baits are to use, treble hooks are difficult to remove from the small mouth of the bluegill. A pair of hemostats or pliers is recommended to remove the hooks.

In-line spinners are effective lures for bluegill, but there are disadvantages. As with other lures such as minnow baits, treble hooks are difficult to remove from the small mouth of a bluegill. Weighted treble hooks are more prone to snagging. In-line spinners are more effective in open water as a search lure. Notice swivel added.

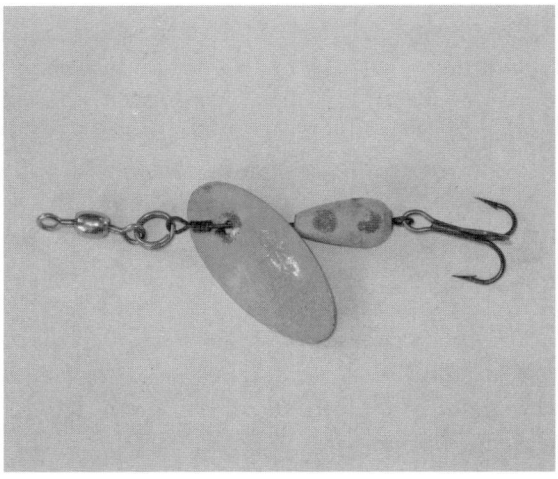

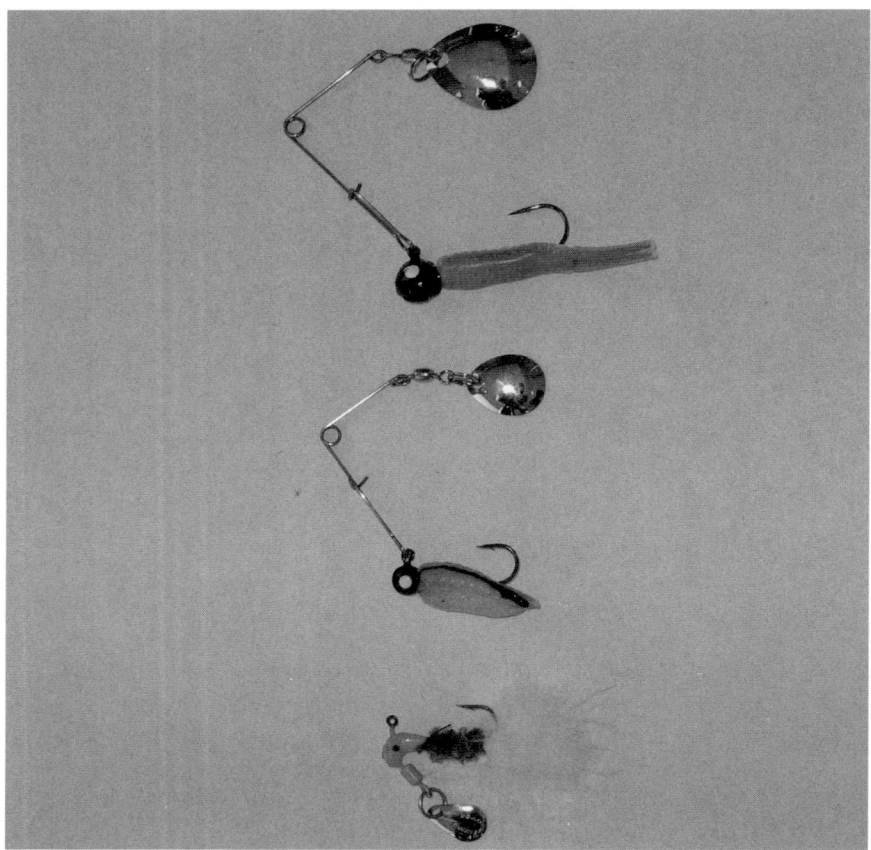

Spinner baits are one of the most versatile baits you can use. Countless numbers and species of fish have been taken with spinner baits. Spinner baits can be fished high in the water column and down on the bottom. Better still, the spinner arm acts as a weed guard, reducing snags. This is a very desirable advantage when fishing wood and vegetation.

The Beetle Spin (top two lures) has been around a long time and is now available with Power Bait grubs, making three dimensions of attraction: flash, sound, and scent. Use one size heavier than you think you need, start with 1/16-oz for 4-pound test line.

Road Runners (bottom) are another spinner bait with a huge following, in particular crappie anglers. The lure is basically the same as the safety-pin style. Available in various dressings or bare so you can add your favorite plastic or live bait.

Jig heads in 1/32-, 1/16-, and 1/8-ounce weights should be adequate for bluegill. Most of the fishing will require 1/32-, 1/16-ounce jigs. Notice the barb on the collar behind the jig head. It's purpose is to keep plastic baits from sliding down the hook shank. A hearty strike will pull the bait down, barb or not.

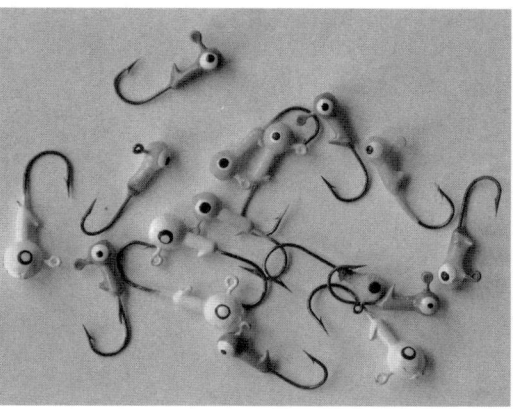

A well-stocked box includes jig heads in various colors and weights; round floats and casting bubbles for fixed-depth presentations; spinner arms to make spinner baits; an eyebuster to clean paint from hook eyes; hook hone to sharpen jig hooks because not all jig hooks come from the package as sharp as is necessary. A couple of large jigs are carried for the rogue bass lurking about. Where you find bluegill, you will find bass. They like the same basic habitat.

You do not need this much tackle to fish for bluegill, but multiple-compartment boxes are convenient in that all of the small pieces can be stored in separate slots for east access. One multiple-compartment box for jigs and lures and a second box for live bait tackle is all you need to be successful.

Jigs are effective when tipped with hair, feathers, synthetic fibers, soft plastics, worms, grubs, leeches, insect larvae, minnows, pieces of fish, and most forms of meat. Most convenient is the jig with fixed dressing, such as the hair or feather jig. Jigs tipped with live bait require a steady supply for continuous changes of bait.

Somewhere between the carefree fixed dressing and live bait is the soft plastic bait. Soft plastics come in every color imaginable, and some are scented. Anglers can often catch more fish with scented plastics than with live bait. But, be forewarned, the scent is developed to attract fish, not people. Pictured above from the top down are examples of Berkley 1" Sparkle Tube (lime green/scales); 1" Sparkle Grub (lime green/scales);

Sparkle Wigglers (chartreuse/scales). The 1" size is ideal for bluegill and they love Power Bait. Lime green is the most effective color in most fishing conditions, but do not limit your choices to one color. Change color if green does not work. Change your presentation before changing color (faster and higher retrieve to slower and lower and vice versa). You will be hard pressed to find more effective jig dressing than 1" Power Bait grubs and tubes in lime green. Why green? It's the vegetation connection.

Remember when using artificial lures: Bluegill are insect eaters, not ambush predators at the top of the food chain. They are capable of catching prey while it is moving, very capable, but bluegill do not eat as the black basses do. You will not see a bluegill leaping into the air. The point is not to rip a jig through the water. Use more of a do-nothing approach; just reel it in slow and steady. You want to stay at about one-half to two-thirds of the depth to the bottom, unless the fish are active, then move up. You are using the jig for its most basic advantage, a weighted hook. The jig does not have to be jigged to be effective when fishing for bluegill.

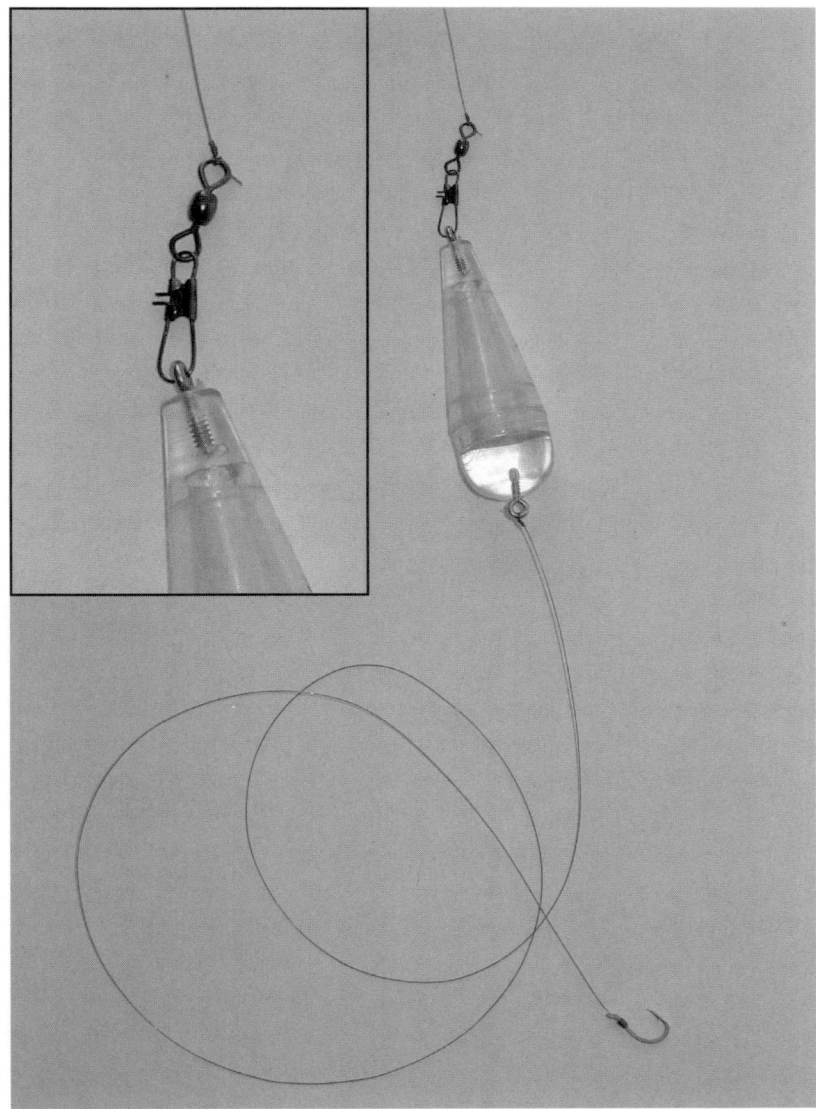

A spin bubble was developed to cast flies with conventional tackle. The bubble works well with bare hooks. Bait the hook with a wax worm, cricket, garden worm, plastic grub, tube or trout fly. The spin bubble is designed as a second, hands-off presentation to accompany your main presentation to deeper water. See also *Chapter 3 for instruction on using this tackle.*

Here is a round float modified to work as a spin bubble. The float is turned upside down. A swivel is attached to the upper end to reduce line twist. Reeling in tackle twists the line; a swivel is a good idea for most rigging. The example above is as effective as it is inexpensive. You do not have to own expensive tackle to catch fish. The foundation of fishing for bluegill consists of bare hooks, floats, weights, and monofilament.

The basic hook styles that anglers will find useful are pictured left to right: Tru-Turn 853 Aberdeen (blood red); Tru-Turn 856 Aberdeen (bronze); Mustad Ultra Point Bait 92569 (fluorescent red); Mustad Ultra Point Demon Circle 39954BLN (nickel). Hooks shown not actual size.

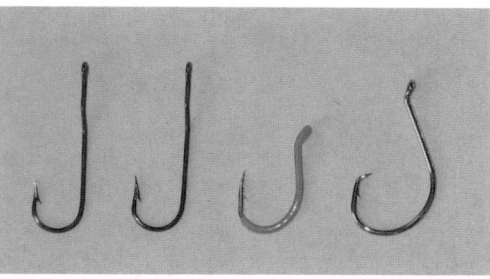

Long, thin Aberdeen hooks are popular with crappie anglers who fish with minnows hooked through the back. Live bait hooks are popular with steelhead and walleye anglers. Both hook styles have advantages. The Tru-Turn design has a kinked shank for increased hooking rates. Well made and extremely effective, Tru-Turn has a loyal following. The Mustad hook has a point that is nearly indestructible and the shape makes it dig in on contact, with your hand if you are not careful.

O. Mustad & Son, Inc. is one of the oldest, most respected hook companies in the world. The Ultra Point series exemplifies the quality and craftsmanship that the most serious anglers demand.

Hooks are a matter of preference. The hooks shown above have been tested and provide excellent results. Some testing has continued for years. But, these are not the only hooks available to the angler. Many styles are available from other companies as well as the TTI Companies and O. Mustad & Son, Inc.

Bronze, blue, black, and nickel finishes work just fine. However, you should give red a try to see for yourself if hook color makes a difference. Red hooks are gaining popularity as an extra visual attraction to the bait. Some think red signifies blood; others think it resembles the exposed gills of a fish. Red hooks are not necessary, but can make a difference.

Hook size for bluegill should be small, size 8 when possible and size 6 if necessary. Hook size for bluegill is often too large. The smaller hook attracts more strikes in pressured water, but will be swallowed more often. The smallest size for circle hooks by Mustad is size 6.

Circle hooks are recommended for all bait fishing. Few, if any fish will be able to swallow a circle hook. The fish you do not want to eat can be released safely. Circle hooks should be considered mandatory when fishing with children and inexperienced anglers. Make sure the circle hooks are not offset; offset circle hooks can be swallowed on a regular basis.

A small box for hooks will keep your hooks separated by size for easy access. Putting hooks back wet or reaching in with wet fingers will cause rust. A hook should be as sharp as possible each time you use it. A rusted hook is no longer as sharp as it should be. If you have wet hooks from a trip, simply open the lid and let dry. Numerous small compartments can store swivels, float stop beads, and split shot with the hooks. Baited hooks are all you need to catch bluegill and lots of them. The bait you choose is important, but not as important as where and when you cast the bait.

Opposite page: *Here are examples of how plastic baits can be presented without having to buy jig heads. Plastic baits hooked as shown are the same way you should hook live bait, once through the end. It does not matter which end; see inset picture of a live grub as an example. Night crawlers present a challenge for bluegill in that they are so large. A bluegill cannot get the whole worm in its mouth and bites off a piece and swims off. Many bites are missed with night crawlers simply because the fish has the worm and is nowhere near the hook.*

Plastic baits hooked as shown attract bluegill when fished soft and slow. Sure there will be times when faster retrieves are attractive, but soft and slow seems irresistible much of the time. You can fish plastic as if live bait, with no motion. The scented plastic and soft texture will do the work for you. If you had hooked the same plastic baits on jigs, you would have had to use a faster presentation to keep the jig from digging into the bottom.

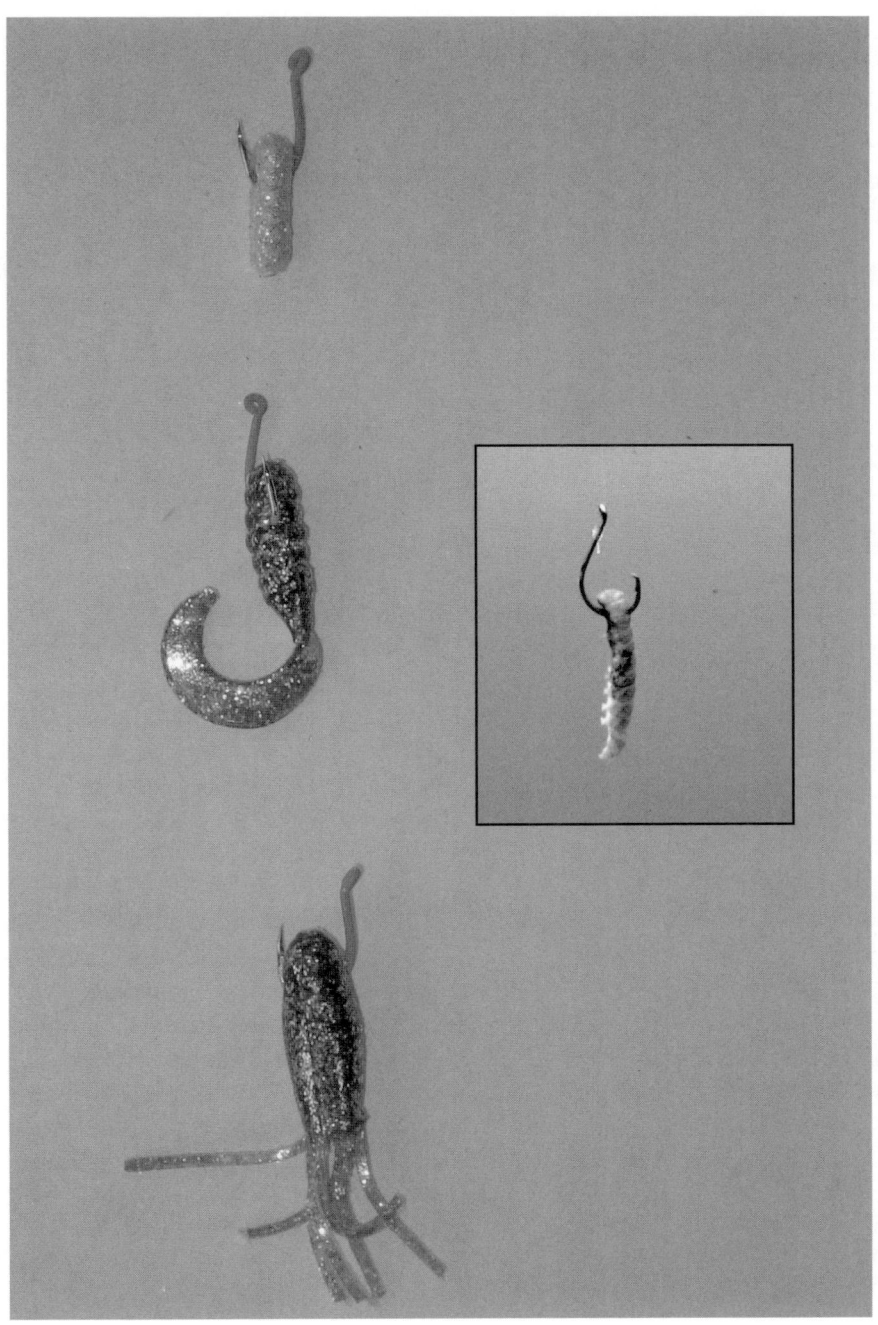

Wax worms make ideal bait for bluegill. Available in bait shops and pet stores, you can catch bluegill under most conditions throughout the year. Ice fishing can be an exception when conditions are tough. Maggots at times work better than wax worms when the water is ice cold. Wax worms shown larger than actual size.

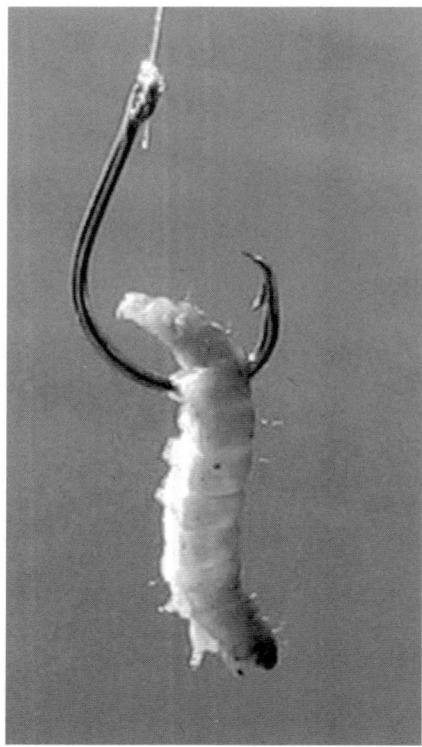

Small hooks should be used with small-sized bait. There is a problem with this strategy, however; small hooks are swallowed when used with small bait. Circle hooks are the answer. Unfortunately circle hooks are not available from Mustad in size 8. Size 8 circle hooks are available from another manufacturer for other species such as trout, but the hooks have offset points and were swallowed when tested. Mustad in size 6 is large for a wax worm, but use the hook anyway, especially with children. Mustad circle hooks are appropriate for all anglers, regardless of skill level.

Worms are perhaps the most popular bait used in freshwater fishing. Night crawlers (left) and garden worms (right) make exceptional bait for bluegill. The garden worm is the preferred bait because of its smaller size. A night crawler is so large that is often bitten off by a bluegill. Pressured fish do not hesitate to take a garden worm, but will refuse a piece of a night crawler after it is broken into pieces or bitten off.

Either worm should be hooked once, as with any bluegill bait, not in the collar as is often done. Hooking through the collar kills the worm. A dead night crawler does not work as well as a live worm. A wriggling worm is irresistible. Hook a night crawler once near the tip or at the middle. Garden worms are hooked near the tip, once. Garden worms are more effective than night crawlers, but are not as available for sale as night crawlers. You can dig your own worms or pick them up from lawns and sidewalks after heavy rainfall. Both types of worms will be out at the same time.

The Trilene Knot

If you learn only one knot, this is the one.

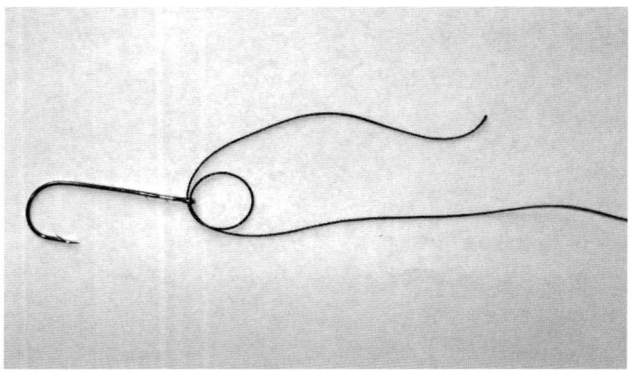

Step 1: Pass the line through the hook eye twice to make a double loop with plenty of extra line to tie with

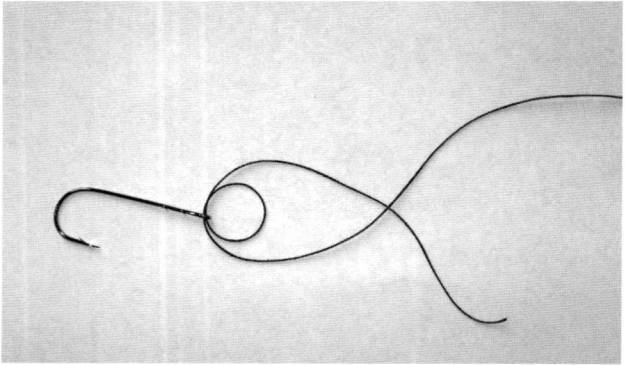

Step 2: Wrap the tag end around the running line

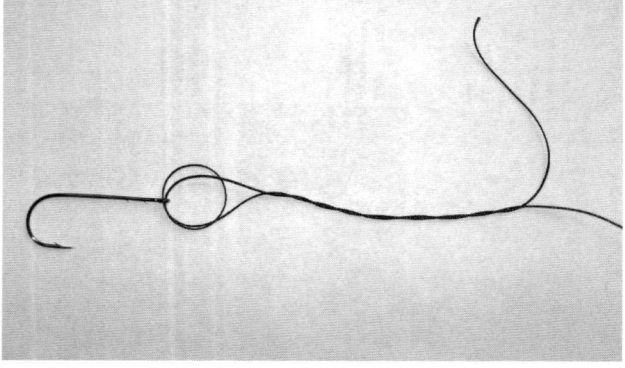

Step 3: Make 5 wraps around the running line

Step 4: Pass the tag end through the double loop

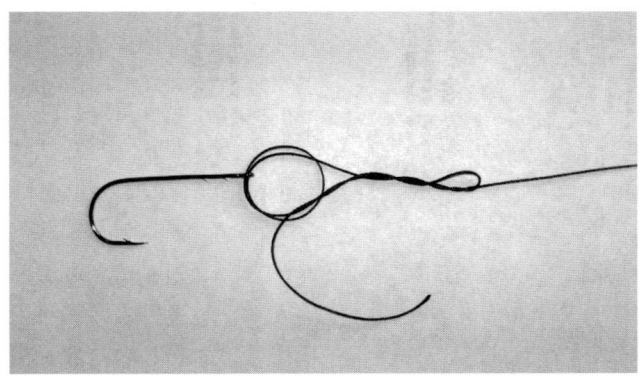

Step 5: Pull the tag end and moisten the knot with water or spit

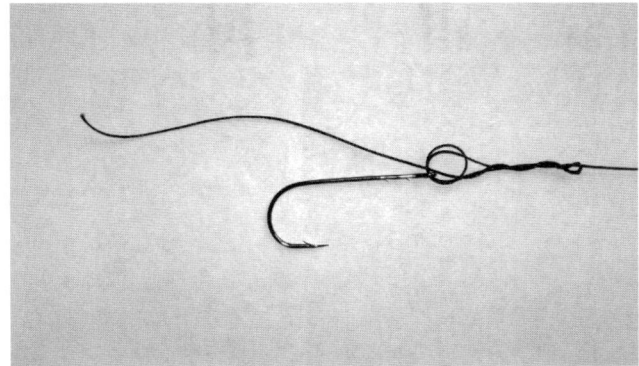

Step 6: Pull the tag end tight with hemostats or pliers and trim

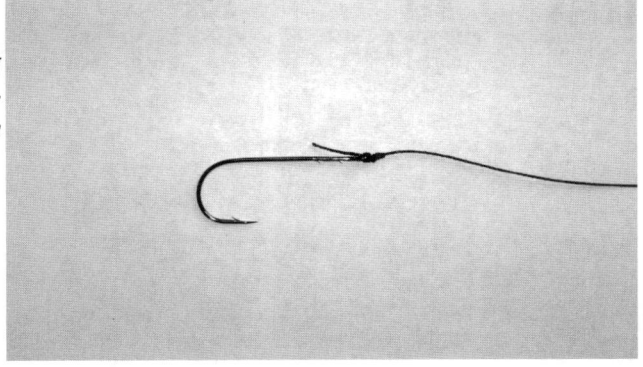

Knots fail when the wraps slip. The Trilene knot is no exception, in particular with 4-pound test. Pulling the tag end tight with a tool is necessary. Berkley recommends trimming the tag end to 1/4 inch, which is not shown to size in the example above.

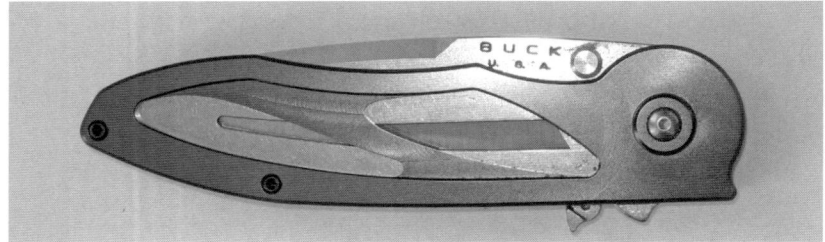

A sharp knife has utility in the outdoors unlike any other tool. When you need a knife, there is no substitute. Keeping a knife on your person or in your tackle box will pay off when you least expect it.

If you are not sure what knife to buy, anything from Buck Knives is a good investment. Choose a pocketknife for general use. One-handed models are preferred because you want a hand free at all times. Choose a sheath knife for specific uses, such as chores back at camp and emergency cutting. Sheath knives have fixed blades and one-handed operation is not the issue, but length can be at times.

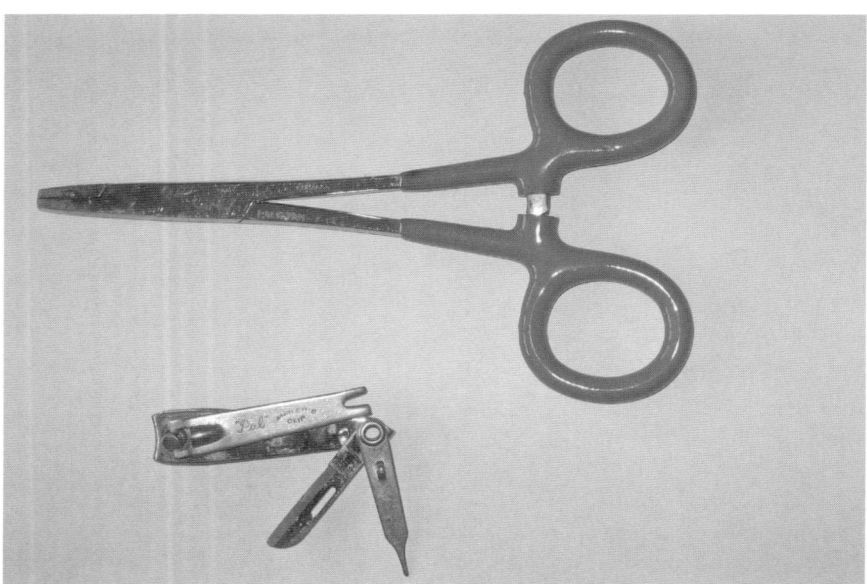

Two tools you cannot do without when fishing are hemostats (above) and an angler tool (below). The angler tool will clip line close to the hook without nicking the running line. Fingernail clippers will do. Hemostats will open and close split shot, pull knots tight, remove hooks from the small mouth of a bluegill, and perform other duties as well.

Minnow baits are not the only option when fishing for bluegill. Insect imitating baits are often more effective than minnow baits because bluegill eat more insects than fish. Rebel makes a wide variety of insect baits such as the Crickhopper shown above. The company also makes minnow baits in sizes small enough for bluegill. This photograph is also the perfect example of the need for hemostats or needle-nosed pliers. Bluegill hit these baits so hard that more than one of the hooks connect, in some cases all three hooks of one treble, as is the case above.

Floating leaf plants grow in shallow water and are not the optimum cover for large bluegill. Better are plants growing completely under water at depths greater than the floating leaf species. But, do not pass up a chance to fish vegetation of any kind, just fish the submerged plants in deeper water, first. Submerged plants grow in the littoral zone. The littoral zone is the area where sunlight reaches the bottom.

3

VEGETATION

How often do you catch big bluegill? Do you know where and when to find them? If you are not satisfied with the size and numbers of bluegill you normally catch, you may not be spending enough time in the vegetation.

PLANT BEDS

Plant beds are the rainforests of lakes and ponds. Submerged plants grow in the areas where sunlight penetrates to the bottom. Clearer water is better because plants will grow in deeper water.

Plants provide food and shelter for numerous species, both predator and prey. Bluegill are drawn to plant beds as staging areas between their deep haunts of winter and their spawning grounds in shallow water.

A plant bed extending down into deep water near an old creek channel or dam provides ideal habitat. Bluegill can move through the water from shallow to deep without leaving the bed. Plants protect bluegill and their young from the wind-induced wave action. The waves from persistent winds shear plants and restrict growth on exposed shores.

Finding a plant bed does not guarantee success. Not just any plant bed will do. A productive plant bed begins with stalked plants. Most people recognize the large round leaves of the pond lily and water lily floating on the surface. A lily normally grows in very shallow water. Submerged plants growing in deeper water are more attractive to bluegill. Look for submerged plants that are growing completely underwater in depths greater than 3 feet.

Curly pondweed is a good example. Native to Europe, but now widespread in the US, each plant arises from a sturdy stalk. The plants grow several inches apart, making travel easier for the fish and easier fishing for the angler.

Slab-sided bluegill can pass between the stalks to escape the maw of a bass or feast on the aquatic insects living among the plants. Other invertebrate species, including crayfish, thrive there as well. A bluegill will eat what it can fit in its mouth.

Often growing among the curly pondweed is native plant commonly called coontail. The plant has no roots; it hangs among the stalks and leaves of other plants.

Another common plant is water milfoil, an invasive species from Europe and Asia. Milfoil has roots and grows in thick stands. With its dense growth and bending stalks, fishing can be tough if the entire plant bed consists of milfoil.

Depending on the structure of a lake or pond, bluegill may use a plant bed for much of the year. But, plant beds are best fished in the spring during the prespawn and spawn when the largest bluegill are in shallow water. Reaching them will be easy if they are in or near the plant beds. Good fishing should last through the spawning season, sometimes longer. Once the majority of the bluegill have spawned, they disperse and are no longer as concentrated as they once were.

By June, many species of submerged plants may have grown too close to the surface for easy fishing. Fishing then becomes restricted to the perimeter of a plant bed and is not always productive. You can reach the bluegill in the openings in the vegetation. A vertical presentation is often necessary. Vertical presentations are most effective with a long pole. And, the temperature may rise above the comfort range of adult bluegill, which will leave for a cooler temperature in deeper water.

Bluegill may spawn over a longer period of time than other species of fish. You may have noticed males guarding nests well into the summer. This leads me to believe that bluegill do not all spawn at the same time. Some males will try to spawn more than once.

Submerged plants are the true magnets for bluegill. Because the plants grow completely underwater they grow at greater depths. Habitat in somewhat deeper water is more attractive. Submerged plants provide habitat not only for bluegill but aquatic insects, crawfishes, and other invertebrates. Habitat and food in deeper water creates an ideal environment for bluegill to thrive. From left to right: curly pondweed, water milfoil, and coontail.

THE SPAWN

You can estimate the progress of a spawning period by the fish caught. If the catch is mostly males, assume the females are still in deep water and the spawn has not begun. If the catch is a somewhat equal number of males and females, the spawn is well on its way. After a week or two of catching both males and females, expect your catches to begin dropping off. The catch again becomes mostly males. The spawning period has peaked. Some of the males are guarding nests and others looking for opportunities to spawn.

THE METHOD

Weeding out bluegill is two-dimensional fishing. The first dimension is fishing with a vertical presentation to the outside edges of a plant bed. The second dimension is fishing with a horizontal presentation over the plant bed. Vertical presentations are aimed at the deep fish, those positioned down among the plant stalks. Horizontal presentations are aimed at the shallow fish close to the surface, those positioned in open pockets of water and among the plant tops.

Fishing over the vegetation can be challenging when the holes are small and the plants are near the surface. If you snag the plants all the time, you are better off fishing deep along the open edges, unless you don't mind the extra effort of pulling free. You will increase your catch by fishing both ways. One method will catch more fish than the other during the course of a day.

Each method is simple but needs specific tackle to be effective. Big bluegill in particular may stop inches away from the bait to watch for a moment. The more fishing pressure put on the bluegill, the more cautious they become. Premium monofilament line is the key to catching big bluegill. Light line presents bait with more natural movement than heavy line and is less likely to spook a wary fish in clear water.

Vertical presentations

A slip float (bobber) was created to present bait in deep water. Rigging for a vertical presentation is simple. A float stop, float, sinker or split shot, swivel, and hook are all you need.

When fishing along a plant bed and especially over one, snags are unavoidable. Poor casting will cause some snagging of your line, unseen plants will cause more, and the bluegill diving for cover will cause the most. Plant snags will not break your line, not even 4-pound test, which is why premium monofilament is necessary.

But, if you use 4-pound test, you should retie the knots frequently, depending on the strain and number of fish caught. Fresh knots will keep a light line strong, but the time needed to retie each knot takes time from fishing. Using a heavier line such as 6- or 8-pound test will reduce the need for fresh knots. Heavier line is the choice if you do not like tying knots. Heavier line is a better choice for anglers who feel more confident with heavier line.

When using 6- or 8-pound test monofilament, casting distance is less than it would be with 4-pound test, so there are positive and negative points about the line weight you use. A presentation with heavier line is stiffer, less natural. Light line is critical when fishing in water open to the public, which receives considerable pressure. Presentations with lighter line are more effective, especially when using artificial baits and lures. More effective presentations catch more fish.

Attach a slip float with buoyancy requiring 1/8- to 1/4-ounce weights for balance. A 1/4-ounce weight is suited to 8-pound test line. Do not use a float with a lead collar. Weighted floats are designed to reduce the number of tangles made while casting.

Wax worms are the bait of choice when fishing for bluegill. You need some weight on the line below the float and about 12 inches above the bait to pull it down to the desired depth because wax worms and other tiny baits used for bluegill fishing are too light to be used without weight added close to the hook.

The best weight to use is a slip sinker. Split shot will work, but pop off when monofilament is stretched. Fishing in the plants is more or less pulling many of the casts free, with or without a fish. The plants give before the line breaks, but pulling on the line stretches it enough to pop the split shot. You will find yourself spending too much time replacing split shot instead of fishing. And, crimping shot numerous times weakens a light line. Rubber core sinkers are an alternative to split shot, but will pop off the same way as split shot.

Below the sinker attach a live-bait or barrel swivel; size 12 will work for 4- to 8-pound test lines. If you prefer a range in sizes, a number 14 is one size smaller and a number 10 is one size larger. Color is not critical; choices are brass, black, and shiny metallic. Black is preferred; it does not take attention away from the bait as a bright finish does.

A swivel reduces line twist. Monofilament twists when reeled in, regardless of what is tied to it, and lighter lines twist more than heavier lines. A swivel also serves as the lower stop for the float and as a link to attach the hook with a short leader made from monofilament.

Start with about an 18-inch length of 4-pound test

monofilament to make a 10-inch leader. If you like plenty of free line to tie knots, start with more than 18 inches of monofilament. You will find the Trilene knot useful for most of your rigging. If the hook has an upturned eye, snell the hook instead of using the Trilene knot.

Tie on a small hook, size 8 or 6, to the leader. The finished leader with a hook attached to the swivel does not have to be 10 inches.

Make float stops (upper-stop knots) from thread used for carpet and buttons; it is heavier than regular sewing thread and available in department stores.

A good depth to set the float is about two-thirds of the depth of the water you are fishing. For example, if you are fishing in 9 feet of water, set the float stop at 6 feet. If you do not get bites at this depth, slide the float stop up or down until you find the fish.

Sometimes watching the float can tell you where the fish are coming from to take the bait. If the float moves straight down, the fish may be deeper. If the float moves up slightly before moving off, the fish may be shallower. If the float moves off in a certain direction, cast in that direction the next time.

You may have read in other publications warnings about fishing deep with a slip float from a long distance away. The conventional wisdom is that the angle is too steep to set the hook. The secret for fishing from long distances is to reel in all the slack before setting the hook. Reel in with a quick, yet smooth motion until you feel the slightest tension of a fish, and then set the hook with a slow, sweeping arc. Better still, use a circle hook; no hookset is necessary. *See also* Chapter 7 for more details on using a circle hook.

Rod choice cannot be neglected if using ultra light line. The thinnest line can be used with excellent results. The key is a long rod having a slow action. Length should be 6'6" or longer.

A greater length gives the angler greater control of the line, which is important when fishing deep water from a distance. Because the rod distributes force over a greater distance, less stress concentrates on the line. A moderate to slow action softens the impact caused by casting the line or setting a hook.

Casting distance is another advantage of a long rod, giving more opportunity to fish from a single location. If you are fishing from shore, you may not be able to move in the first place. The benefit is spending more time with bait in the water, instead of walking around on shore, looking for a clear place to launch the next cast. You will find yourself doing the same to achieve distance on the water: moving the boat, anchoring, rebaiting, etc.

Big bluegill are hard to come by; you are not going to catch one on every cast, nor will you be able to catch them all from one location. Distance is an advantage no angler can ignore. A long rod strung with light line casts well against the wind, too.

Horizontal presentations

A horizontal presentation to the surface can be as effective, if not more effective, than a vertical presentation to deep water. Used together, the combined presentations can be adapted for just about any bluegill habitat, not just vegetation.

Attach by pushing the button on top of the float as if attaching to your line. Then, clip the rounded end (bottom) of the float to the lower eye of the swivel. Correctly attached, the bottom of the float is up toward the rod tip and the button is pointed downward.

To the bottom clip of the float, add a leader, 3 feet or so, made from premium 4-pound test monofilament. Then, tie on the same hook as in the first method.

Horizontal presentations require a 6' or 6'6" rod. You will need to make accurate casts to the pockets and this may be easier with a standard-length rod. As with the vertical method,

Plant beds are the rain forests of a lake. Notice the plants growing to the surface. Notice even more the open water between the dam and the plants growing to the surface. You could not ask for better bluegill habitat: plant beds growing in deep water. The vertical and horizontal presentations prove valuable in such places because bluegill will move around through the course of a day.

use a premium brand of 4- to 8-pound test monofilament as the running line. Tie on a swivel matching the line weight to reduce line twist and create a link to attach a round float (plastic push-button style).

Choose a float about 1¼ inch in diameter to provide the weight needed to cast without adding a sinker or split shot. A yellow and orange float is a good choice for visibility.

No matter what colors you choose, try to avoid a float with a white bottom. The white will be under water and visible to the fish. Bluegill seem to shy away from white-bottomed floats when they are not in an active mood to feed. This shyness will not happen all of the time, but enough for you to notice a difference in strikes when using two different floats side-by-side. Panfish floats are often white on the bottom, not the classic round float, which is normally white on top for greater visibility.

A clear-plastic spin bubble is another option to use. But, there are issues you should know about. First, the float can be very heavy for ultra light tackle. Second, the float is clear and invisible at a distance. Third, the bass won't leave the clear float alone. This is hardly a problem, unless the bass scare the bluegill away.

Do not add a sinker or split shot to the line or leader when fishing over the vegetation. You want the bait to drift free among the plant tops.

Once a fish is hooked over vegetation, reel it in fast to keep it from burrowing into the mass of vegetation. If it burrows in, the fight is over and the fish must be pulled out, along with a couple of pounds of vegetation. Lightly hooked fish may be pulled off.

An adult bluegill pulls hard; it isn't reeled in without a fight, especially in dense vegetation, which is why 8-pound monofilament may be more appropriate as a running line. Use 8-pound test if the 6-pound feels like it is stretching when you pull hard.

You will have great success catching bluegill with light line, but if you have pulled hard on a hung fish or snag, replace the leader if using 4- and 6-pound test monofilament and retie the knot in the running line above the swivel. The knots are the weakest points when fishing around vegetation. Rocks and other hard objects that can ruin light line are scraping and weakening the line.

TAKING AIM

Accurate casting is important when fishing around plant beds, even when fishing for panfish with live bait. Fish both deep and shallow at the same time using the vertical and horizontal methods.

Start by casting parallel to the deep edge of the plant bed. Align the boat to fish the edge in front and behind the boat without moving. Then fan cast to the open pockets over the plant bed with the other rod. If you are fishing from the bank, line up your position as best you can. A clear shot to the plant bed is not guaranteed when fishing from shore. You may not be able to use both presentations from a single location.

Cast the vertical rig as close as possible to the edge of the vegetation and set the rod down. Polarized glasses will help you to see deep into the water. Cast the horizontal rig over the plant bed then set it down. Sit back and relax for a few minutes while you watch for a strike.

Let each rod sit for about 3 minutes. Then pick up the vertical rig and cast it about 10 feet farther than the previous cast. Then do the same with the horizontal rig, but casting it to the left or right as well as farther out. Repeat this process until you have searched the water within casting distance.

Crank the reel a few turns before reeling in for the next cast. Big bluegill are selective feeders, not ambush predators. They begin their lives eating microscopic prey and are conditioned to examine closely what they are about to eat. Don't be surprised by a strike as soon as you move the bait a foot or two. Moving the bait a short distance often triggers a strike if a fish has been watching.

If you like big bluegill, you will have to work for them. Simply casting to the plant bed is not going to fill your basket. Pick each cast, aim for the irregular features such as points, corners, and pockets. A large adult claims a unique feature in the bed. Fishing around vegetation is not easy; it requires attention to detail, but that is how big bluegill live, with attention to detail.

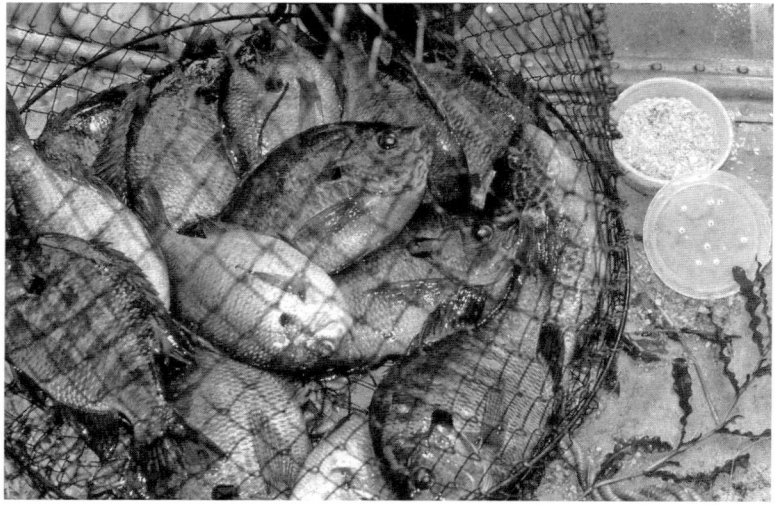

78 • CATCHING BLUEGILL

Electronic depth finders are not necessary to find bluegill, but can certainly help if you know how to use these devices.

4

FINDING FISH

NORMAL CONDITIONS

Finding fish under normal conditions will depend on how well you understand how a bluegill lives. Finding is more difficult than catching. Consistent success catching any species of fish requires a clear understanding of its behavior.

Bluegill behavior focuses on food sources and areas providing shelter near the source of food. Spawning is a major event that affects bluegill behavior, but it is short term. Depending on the character of the lake or pond, spawning activities may cause the bluegill to congregate near shore areas for two or three weeks. Some years it will be less than this and other years more.

Temperature and rainfall will affect the spawning activities. Where they go after spawning leads to deeper water and food. The exact locations a bluegill chooses will change not only through the course of a season, as might be expected, but through the course of a day as well.

Temperature affects the location of a bluegill after it has spawned. As the temperature drops so do the fish; they go deeper in the water. Deeper in the water does not mean leaving the general area, rather going down slope of their present location.

Sunlight has the opposite effect. As the sun rises, the fish descend. When the sun sets, the fish rise. Often they swim toward shallow water by rising to the surface or swimming toward shore.

FOOD

Food is important when it comes to finding bluegill. The location of preferred food determines the general location of the bluegill. Because zooplankton, insects, aquatic worms, crayfish, and other invertebrates make up the largest component of a typical bluegill diet, the bluegill is not restricted to specific structures. Such a wide range of forage can be found in numerous places within a lake or pond, including open water not associated with plant beds, rocks, and timber. Plankton for example drifts freely in the open water.

Aquatic insects

Anglers who fly-fish in streams understand the importance of the aquatic insect communities. Lakes and ponds have their own insect communities. The species of insects living in a stream are not the same as the lake species. Each lives in a different environment.

As stated in Chapter 3, a plant bed is the rainforest of a body of water. But aquatic insects live in many areas not associated with vegetation. Tree roots and deadfalls support numerous species, as do rocks and gravel. Some species, such as the mosquito, drift in the water column. Other insects live in the beds of sand settled on the bottom. If rocks are on the bottom, less space is available for some of the insect species preferring sand beds, but more habitats are available for species that seek shelter among the rocks. This includes predator and prey species.

A good example of an important insect species is the burrowing mayfly. Called the willow fly in the South, this larva grows to a large size as far as insects go, and it gets the undivided attention of the bluegill population when it emerges from the lake bed to hatch. Insect hatches attract the attention of numerous species of fish, not just the bluegill.

Burrowing mayflies need clean water and sand to thrive. Clean sand settles in flat areas of the lake after strong winds have mixed the water. Tributaries also mix the water after heavy rainfall and snowmelt.

Midges are other species of insects that thrive in sand beds. These relatives of the mosquito live in densely populated communities, normally greater in number than the other groups of insects. Though tiny, they make ideal forage for the bluegill.

Crayfish

Crayfish provide a tremendous forage base for many species of fish, not just bluegill. The amount of crayfish in the bluegill diet is not as great as other sunfish such as the rock bass and smallmouth bass. Predation of crayfish depends on the habitats of the fish species in question. The way a fish feeds—bottom feeder, through the column feeder—determines how many crayfish it normally eats.

An average bluegill does not have a mouth large enough to eat an adult crayfish. Juvenile crayfish are fair game. The young crayfish leaving their mother to find a place of their own are preyed upon heavily by numerous species of fish. Crayfish in the bluegill diet is common in the spring and summer when the crayfish are hatching and dispersing to new territories. Some crayfish will hatch in the fall, but much of the activity occurs in the spring and early summer when the water temperature is rising.

Fish

Fish eggs and fry supplement the diets of many species of fish. But bluegill are notorious nest robbers, first eating the eggs, then the fry.

Male largemouth bass have to guard a nest from bluegill waiting nearby for a chance to swim in and eat the eggs and fry. If the male bass abandons a nest for a short period of time, the bluegill nearby will eat all of the eggs and fry. Few, if any, bass will survive in an unguarded nest.

Newly hatched fish are most vulnerable just after leaving the nest. These young-of-the-year fish are preyed upon by bluegill until the fry grow beyond the gape of a bluegill. Minnows do not attain a size large enough to avoid predation by adult bluegill.

Finding bluegill in the spring may be as easy as finding the spawning and nursery grounds of largemouth bass, which is the same or similar to the bluegill. Largemouth bass spawn in temperatures ranging from 60-70 degrees Fahrenheit, concurrent with the spawning temperatures of bluegill.

HABITAT

If you can identify habitat located close to food sources, you find an ideal place to catch bluegill. What is habitat? A change in structure of the bottom, any object(s) rising from the bottom, floating on the surface, and suspended in the water; most materials, whether natural or manufactured, will create viable habitats for a bluegill.

Habitat attracts not because of the protection it offers; the sense of security is what counts. Habitat can span from one of extreme protection, such as a fortress of plants growing in a dense bed, to one of minimal protection, such as the shadow cast by a floating dock. Both forms attract bluegill. What makes one form of habitat more or less attractive is the closeness to food. The closeness to deep water is also important to the largest bluegill. This means habitat located near food and deep water has great potential to attract bluegill.

Habitat can be one-sided, attracting the predator or the prey. Some forms will attract the predator and their prey if food and shelter are available to both.

Habitat for the prey is a given. But, what we recognize as habitat is not always what it appears to be, in particular when considering habitat for the prey of the bluegill. Bluegill eat invertebrates and invertebrates live throughout, so finding some form of prey is not as challenging as it would be for a top predator such as a northern pike.

Open water

Some of the most enjoyable fishing can be found in open water. Enjoyable because the bluegill are often near the surface, away from the snags that foul a line or snag a hook. Bluegill at times suspend over deep water to feed on plankton and insect hatches. Because they are there for food and close to the surface, topwater presentations are fun and effective. But, bluegill swimming out in the open water is not as common as one might think.

Finding bluegill in open water is not easy; during the heat of the

Fishing in a lake is no different than fishing in a pond other than the size of the body of water. Bluegill are not scattered throughout every inch of a pond and the same is true of bluegill in a lake.

day is the toughest time. Bluegill will not often be exposed in direct sunlight without cover around them. You may think the bluegill will be everywhere; they're bluegill, right? Wait until you try to catch them with that thought in mind. You can't just put on bait, add water, and expect good results. Simply casting to open water may work on occasion, but not often enough to keep most anglers happy, especially young anglers.

Bluegill feel vulnerable in the open. Traveling in loose schools offers some form of protection and security. That said, a heavy splashdown from a lure and even a fly laid with a careless cast easily spooks bluegill away from the security of deep water or vegetation, especially the largest fish. Fly-fishing is not as much an issue as the much heavier baits used with spinning and casting tackle.

How active the bluegill are and how much fishing pressure they receive determines how sensitive the bluegill will be to disturbances. When the bluegill are actively feeding, lesser disturbances attract their attention as a possible prey item. Greater disturbances can make them flee for cover as if a predator is in the area feeding. A delicate presentation is best under most circumstances, simply because bluegill eat small prey. What they are used to eating does not land like a rock.

Flats

The bottom of a typical lake is flat rather than made of rolling hills. Flats can be covered with numerous forms of habitat. What habitat is present depends on many factors. Geology for one will make a difference as to how the bottom is structured with rocks in different shapes and sizes. All dimensions, from boulder to sand grain may be present, and each contributes to some form of habitat. Boulders provide shelter for the predators and sand grains provides shelter for the prey.

Flats are depositional areas where sand grains and other small-sized materials settle. If water flows over a flat leading to the dam, the movement keeps the flats clean. Flats exposed to the wind are swept clean by wave action.

Mucked flats are backwater areas lacking direct contact with flow and wave action. These areas become stagnant with fine-grained particles such as clay and silt. The particles settle in deep layers.

Many species of insect larvae spend the aquatic portion of their life cycle in the sediments. The diversity of insect larvae depends on the lake bottom. More diversity can be expected in clean sand and less in muck.

Bluegill feed heavily on insect larvae during the warmer months but less heavily during the colder months. Insect larvae are not as available during the colder months because most of the insect species have dug into the bottom for the winter. Some spend the winter as eggs.

Flats attract fish in places where the bottom structure changes, especially those areas where clean sand meets rock. Clean sand attracts aquatic insects. Rocks attract bluegill looking for secure habitat close to a source of food. And, additional sources of food can be found living among the rocks, making rocks adjacent to sand very attractive

Flats are easy to find and mark in the autumn when reservoirs are drawn down to winter pool. Note the lack of rocks and wood.

Points and humps

Points and humps represent the traditional habitat of bass and walleye. And for good reason. A raised formation on an otherwise flat terrain is like a chair in an empty room—inviting. Points and humps are natural attractions for bluegill as well, places to orient to and rest. A raised formation piled with rocks provides more diverse habitat than a raised formation layered with sand.

The attraction of a raised formation with rocks is twofold. The formation is surrounded by deeper water and the rocks provide endless cavities in which the bluegill may find food and shelter. Deeper water means an escape route for the bluegill if needed. Another benefit is the travel corridor provided by the raised landmark, a path to follow between shallow and deep water.

Channels and troughs

Channels and troughs play a similar role as points and humps, but in reverse; the formations are sunken below the surround surfaces. The attraction is the break in the topography, a break that leads to deeper water. As expected, rocks increase attractiveness.

Low water is the best time to look for structures on the bottom. A lake is flat for the most part; it is the slight changes in the bottom elevation that attract bluegill, as well as other species. The change in depth does not have to be great, but every little bit helps. A point is as important as an old channel.

Steep banks

A bank that drops straight down can be hit or miss in terms of providing habitat suitable for bluegill. A steep bank leads to deep water, which is good. But, deep water alone is no guarantee bluegill will be attracted.

The face of the bank is important. If it is a smooth cliff, then it is less attractive than if objects are jutting out below the water line, objects such as clay deposits, tree roots, slabs of rock, and concrete. Overhanging objects provide a shaded refuge. Fallen rocks piled up at the base of the bank are also attractive to bluegill.

Flats attract bluegill when plant beds, wood, rocks, points and humps, and old creek channels are present. If the bank looks clean and flat, you are apt to find more bluegill at a steeper bank or at least one with more habitat at the edge of water. If you cannot see bluegill habitat from where you are standing, and you are standing on a flat bank, you would be better off moving and trying to find a place with more diversity. If you cannot see diversity in the water, you should see trees and rocks on the bank above the water line; anything other than a flat bank completely bare as shown in the picture to the right.

Here is what you want to see on a bank: Steepness to the water, wood everywhere, boulders, brush growing on the bank. The picture above shows the bank during a low-water period, but these observations remain true for water at a normal level. Look for diversity in habitat when plant beds are not visible.

TACTICS

Fishing for bluegill in the open water does not have to be difficult. Using electronics to find bluegill is an option, but looking for visual clues is a simple yet effective way to catch them. I prefer to rely on my senses rather than radar technology when fishing for bluegill. With practice you will recognize opportunities at a glance. When you catch fish, examine the area to learn why the bluegill are there and what tactics work best when they are.

Obvious attractions

Fish in the most obvious places, first. Most obvious are those you can see. The beauty of fishing for bluegill is the simplicity of the tactics and the tackle. You can rely on visible clues, which is effective much of the time.

The most important visible clue is where sunlight reaches the bottom, defined earlier as the littoral zone. The width and depth of the littoral zone determines the productivity of the water. More sunlight reaching the bottom makes the water more productive. The food chain (forage base) is determined by the amount of sunlight—the primary source of energy—entering the depths. The exact depth penetrated depends on water clarity. For example, 8 feet may be the maximum depth in muddy (turbid) water and 15 feet or more in clear water.

Vegetation attracts bluegill more than other forms of habitat. But, bluegill living in some lakes and ponds are going to leave the weeds after spawning.

As a good rule, begin looking for bluegill in the vegetation. Spawning activities will dictate where the bluegill should be in relation to the plant beds. Before spawning, the bluegill will be looking for spawning grounds and feeding in shallow water.

After spawning, a portion of the bluegill will migrate toward deeper water. If the plant beds are extensive, growing down into deep water, some of the bluegill take positions at the deep edges and in holes among the plants. Vegetation growing on points and humps provides ideal habitat for bluegill.

But if you cannot see plants growing, look for other forms of habitat in the areas where sunlight is reaching the bottom. Plants may or may not be growing there, but if sunlight penetrates the water, plankton will be there.

Any form of habitat offering shelter such as wood and rocks will be desirable if located in the littoral zone. When in doubt about the depth, start shallow and work your way to deeper water. Do not go too deep, thinking you will eventually find them. Move to another location closer to shore rather than continuing away from shore.

Wood

Standing timber and fallen trees may be the next best habitat to try after vegetation. Wood, dead or alive, attracts bluegill, but rarely as well as vegetation. Wood becomes attractive when vegetation is not growing in water where other favorable conditions such as depth, temperature, and shade exist around the wood. Normally this occurs when vegetation is absent.

Wood sticking up out of the water is an obvious sign. Wood rising close to the surface is not obvious unless the water is clear and you are close to the wood.

If you have wood everywhere, then you need to choose where to fish. You will not catch bluegill in every stickup or piece of wood you see. When you have choices, look for the wood that is different

Small branches are more attractive than a bare log. Plants add even more attractiveness. Notice how close the float is to the wood.

from the rest. Yes, success is in the details, even with a stick of wood.

What makes one piece of wood better than another? Location is a major factor. Wood located close to deep water is a perfect example. If one factor can be credited with increasing attraction to fish, not just bluegill, it is deep water. Deep water offers security; it does not have to be a sheer drop off to attract fish, just deeper than the surrounding area, even if only a couple of feet. A bluegill does not want to be spotted in places surrounded by shallow water, out in the open with no cover in sight.

What do you do when all the wood is in a similar location? Then, you look for subtle differences, such as how close the wood is to the bank. Wood sticking up in a plant bed is ideal because two types of cover are provided, making the habitat more diverse. But, these areas will be difficult to fish; the more difficult to fish, the more attractive the habitat is to the bluegill during the daylight hours. Early in the morning and late in the evening, bluegill move around to feed.

These details may seem trivial, but you are looking for areas that will receive more current, so to speak. A lake does not have current in the same degree as a flowing stream, but water does flow toward the dam through the old creek channels that once flowed before impoundment. Plankton follows the currents and so do the food chains.

Wave action is another consideration. The wind-induced waves blow along the contours of the lake basin. Wave action can restrict and inhibit plant growth. The lakeshores receiving the most consistent wave action may not provide the most desirable habitat. Areas adjacent to the wave-swept shores may prove the more attractive places to find bluegill. But, during cold periods and low-light conditions, wave-swept shores will have plankton piled up and the bluegill will be feeding there. A general rule concerning habitat: Areas that are productive during pleasant conditions may not be as good when conditions are unpleasant for the angler, locations change with the weather.

Last but not least is the individual piece of wood you intend to fish. More branches create more habitat. If the tree has many fine branches still attached, the attraction to bluegill is greater. An old log that has lost branches and most limbs does not attract as many fish as the tree with many branches. Remember, the more difficult to fish, the more attractive the habitat.

But, the size of the log, regardless of remaining limbs and branches, may offer a few trophy fish what they want in habitat. To be sure, fish the wood next to deep water or main channel areas first—especially wood resting on points. Fish your way back and forth along a point, from shallow to deep or deep to shallow.

If the timber is still alive and standing, fine-stemmed roots provide outstanding habitat for prey. Crayfish, minnows and small fish, aquatic beetles, damselfly larvae, and numerous other prey items are living among the roots. Plus, the roots provide shelter for the bluegill. Overhanging limbs with leaves attached provide shade and the occasional terrestrial insect that falls to the water.

Rocks

Rocks attract bluegill. Look for rocks lining the dam and banks. In suburban ponds, rocky inlets with flowing water are ideal areas to fish.

Rocky mounds and points attract many species of fish, including bluegill. You may have noticed while fishing for bass or walleye, that big bluegill were also present.

Just like other forms of habitat, rocks on slopes (points, humps, banks, and the dam) surrounded by deeper water attract bluegill more than habitat on flat bottoms. The rocky slopes may not provide enough habitat for all the bluegill in the lake. Some of the population, the younger fish in particular, will have to settle for what is left, such as habitat on flatter bottom and in shallower water. Fish the rocks and other forms of habitat on sloped areas first, then try the flats.

Bluegill will be on the flats all right, but finding them will be more difficult unless you look for the places where two forms of habitat come together, such as rocks and clean sand or plants growing near wood.

Docks

Bluegill love docks. Shadows create dark places to hide and protection from direct sunlight. Posts and supports under the docks provide physical features, man-made habitat. Habitat in the shade is more attractive than shade alone and habitat exposed to direct sunlight.

Docks with brush underneath are especially attractive. Anglers who know this throw brush at the ends of the dock, but unless you own the dock, make sure you have permission to do so.

Brush and other objects can interfere with boat traffic, the main purpose for the dock. If you throw brush off a dock, place the brush at the very end where boats do not travel, not on the sides. The inside corners beyond where boats will come is another consideration. Loose brush will move with the wave action.

Docks constructed along windswept shores may prove impossible to seed with brush, no matter how much weight is used to anchor the piles down. The brush will end in the corners where debris is naturally collecting, no need to add more. You can always make your own brush piles in protected coves, to everyone else's delight.

Docks can be magnets for bluegill, but as with any other form of habitat, the fish will not be just anywhere. Fish docks with brush; first (opposite page, top); then try corners, as close as possible to take advantage of the shade (opposite page, bottom); then try the docks farthest out; if these places do not hold fish, try the docks closest to shore. Do not spend time at each place; the bites will be immediate or not at all. If you do not have wax worms or garden worms, dig some grubs from the woods. Grubs can be found under the bark of dead trees and logs. These grubs (inset) were pulled from wild black cherry logs in the campground.

SUSPENDED FISH

Bluegill suspend for the same reasons largemouth bass do, mostly food and water quality. Suspended bluegill are fun to catch when you can find them. Fun because they are often close to the surface following prey. Suspend sounds like a permanent location, but think of it as a temporary position to gain an advantage.

Bluegill that suspend for water quality reasons are not so easy to catch. They may be stressed and not in a feeding mood. Fish do not eat when under stress. Unlike humans, bluegill do not eat to calm their nerves or soothe their feelings.

Schools of bluegill often suspend over deep water to feed on numerous species of zooplankton such as *Daphnia* spp. Zooplankton are a major source of food for bluegill, especially when the fish are small. Even as the bluegill grows to adult size, the important prey items continue to be zooplankton and aquatic insects.

Bluegill readily eat fish, but their small mouth restricts their ability to eat larger-sized prey such as minnows. On the other hand, bluegill love fish eggs and newly hatched fry. A few bluegill can destroy a nest of eggs in minutes if the nest is not guarded.

The movement of suspended fish coincides with the availability of their prey. Bluegill take positions with intent to feed. They look for areas of opportunity. Ideal locations conserve energy when the bluegill feed because they do not have to travel far or hunt to find food.

How and when bluegill suspend in open water is not the same for all bodies of water, nor will it be the same within a body of water. The locations where bluegill spend time depend on the amounts and types of habitat as well as the prey available. The habitat varies with the weather and climate, and the prey varies with the season.

Some bluegill will stay in deep water much of the year; this is common behavior among the largest fish. Deep water for some bluegill may be 15 feet or more in depth. Other bluegill living in the same lake will choose shallow water, say 12 feet or less.

The deep-water bluegill eat what is found close to their locations. The shallow-living bluegill have their own choices for prey as well. The bluegill living in shallow water will have more choices in prey because of the prominent littoral zone and greater diversity in habitat. Large adults move shallow to feed if they are not finding enough food in deep water.

Active fish

This is an ideal time to fish with surface presentations. Bluegill looking for food will charge a surface disturbance. Casting just about any surface lure, popper, or plug will bring results. Be ready with small lures and baits for the times when bluegill are hunting prey and willing to strike anything that resembles food.

Just before sunset is an active time for feeding. You can consistently find bluegill rising close to the surface and moving toward the shore as the sun begins to set. Plankton rise toward the surface as the light intensity decreases. Sunrise is also an effective time, but it may be easier to be on the water at sunset rather than sunrise.

Inactive or neutral fish

Bluegill not actively feeding are a tough challenge, especially if the fish are suspended over deep water away from vegetation or bottom structure. Inactive fish are resting and largely not interested in making an effort to eat. Neutral fish are somewhere in between active and inactive. More often than not, you will find the bluegill in a neutral mood. Let's face it, who can only fish at the most optimum times?

Both inactive and neutral fish can be caught, if you make it effortless for them to eat. Put live bait in their face and they'll bite. Artificial bait presented the same way may be effective for neutral fish, but not necessarily inactive fish.

Scented soft baits such as plastic grubs can catch bluegill when the fishing is slow, but live bait is tough to beat when the bluegill are not active. And, no matter what bait is used, the bait presentation must be effective, live bait or not. You have to use a deliberate approach, a slow or motionless in-their-face offer that cannot be refused because no effort is needed to eat.

PRESENTATIONS

When casting to inactive or neutral fish, the disturbance may scatter them for a few minutes or the duration of your trip if the disturbance is great enough. Active fish react positively to the disturbance.

What is a disturbance great enough? Several factors determine how spooked the bluegill will be. Fishing pressure can affect how the bluegill react to a splashdown. Heavy pressure in popular lakes can encourage the fish to avoid a familiar disturbance. And, the older the fish, the wiser, even in private waters receiving little pressure. Trophy bluegill disappear after the slightest disturbance.

Position in the water and water depth also contribute to how a bluegill will react to bait splashing down. Fish near the surface will spook easier than those positioned down deep. Bluegill in shallow water will spook easier than fish in deep water.

Feeding mood is the final variable that can determine the bluegill reactions to the conditions described above. If you are fishing at sunrise or sunset, chances are the feeding mood is positive and splashdowns will not be a problem. Watch the reaction around the lure when it lands in the water. If bluegill scatter, cast way beyond your target (the area suspected to hold fish). Then slowly bring the lure to the spot. Keep in mind a lighter lure or floating bait is appropriate when bluegill are suspended near the surface or feeding in shallow water.

Flies make ideal baits for suspended bluegill, spooky or not. Flies land with a more subtle impact. The impact from a fly presented properly is so subtle that it attracts more fish than it scares. The first impression is of a real insect alighting on the water.

Regardless of the tackle used, a soft landing will attract more attention because the normal prey eaten makes little, if any, impact on the surface. Remember, bluegill eat small prey. Many of these prey items are too small to see without a microscope.

RIVERS AND STREAMS

Bluegill live in rivers and streams, but you will not find them in the current such as pictured above. They will instead live in the backwater areas not exposed to the direct current.

Water moving in a lake or pond is pleasant breeze, not flowing, but not standing still, either. Rivers and streams are a different story. Rivers and streams flow like the wind. Bluegill do not like flowing water, they are not designed for long periods in the current. You can catch bluegill in all forms of flowing water, from tiny streams to large rivers, but their habitat needs are more closely related to the largemouth bass than the smallmouth bass.

Still water is the key to finding bluegill in rivers and streams. Backwater areas are ideal places to fish. A backwater area does not become stagnant as it can in a lake. Each time it rains enough to raise the river, the high water flushes the backwater.

The backwater is calm because the water drifts away from the main thrust of flow. Bluegill select habitat in rivers and streams the same as if they were living in a lake or pond. Vegetation, wood, and rocks, the desirable forms of habitat in lakes and ponds, are just as attractive in stillwater areas of rivers and streams. The tactics used in lakes and ponds should work equally well in rivers and streams.

Main channel areas can also be productive if the water is calm. Look for familiar forms of habitat and fish accordingly. Don't be surprised if you catch other species of sunfish more often than bluegill when fishing a section of the main channel.

When fishing in rivers and streams, get away from the full thrust of the flow to find calm places and backwater areas. Then look for the same habitat as if you were fishing in a lake. Calm areas of rivers and streams often support bluegill in abundant numbers.

5

FLY TACKLE

Fishing for a bluegill is similar to fishing for a native trout in that delicate presentations make a difference in catch rates. The most delicate presentations prove more effective than coarse landings. Ultra light tackle means extreme action, but be prepared for disadvantages to accompany the obvious advantages of using tackle that is 3-weight or lighter.

FLY LINE

Fishing equipment should be based on the mass of the fly to be used. Mass is the resistance of the fly to accelerate as it is pulled through the air by the line. Choose equipment beginning with a line heavy enough to overcome the mass of the fly. A line of proper weight line leads the fly back and forth through the air with smooth motions. An underweight line does not absorb enough energy to carry the fly through the motions. I feel the action is forced jerking more than balanced casting.

    A fly adds a payload to the line. Each line has a limit to the payload it can carry and still function as designed. Performance from a fly line varies with the changes in the diameter of the line—how the line increases and decreases through the tapers and the lengths of each section (tip, front taper, belly, rear taper, and running line).

    A rod imparts kinetic energy to the mass of the line; it is the line that pulls the fly back and forth. If the line is light in weight and the fly is heavy or bulky (air resistant), then the energy is unevenly distributed between the two masses. The result is less efficient than if the line absorbs the bulk of energy and the fly goes along for the ride.

When choosing the proper line, the shape of a fly line is as important as the weight. The shape determines how a line travels through the air and how it handles in the water.

Line stiffness is also a design trait. Stiffness comes from the core material used in making the line. In cold water, a line designed for less stiffness (low memory) is desirable because the cold temperature stiffens the line. In warm water, a line designed with more stiffness is desirable because the line would otherwise sag between the guides, increasing friction. Friction is undesirable because an increase in friction causes a decrease in casting distance.

SECTIONS OF A LINE

The head is the thick part of the line—what gives a line its shape. The head can be one sided, such as in a weight forward line, or continuous, as in a double taper line. Also called tapers, heads are designed to fulfill one major function such as power, distance, finesse, or general purpose.

Contained in the standard head are the tip, front taper, belly, and rear taper. The belly is the thickest part of the line. The length of a belly determines the amount of control an angler has on the line. The thickness of the line through its various sections influences the performance in the air and in the water. The following is a detailed explanation of the sections of a standard fly line.

Tip

A short section of thin diameter, the tip allows the angler to replace the leader without cutting into the front taper and changing the performance of the line.

Front taper

The beginning of the head, the front taper provides a transition between the thin tip and the thick belly. The angle of the front taper affects how a fly is delivered. Shallow tapers are longer. A longer front taper produces a more delicate presentation as energy dissipates over a longer distance before reaching the leader. A shorter front taper produces a powerful turnover because less energy dissipates.

Belly

The section of line with the greatest diameter, the belly carries much of the energy transferred from the rod. A longer belly improves casting distance and accuracy. A shorter belly is easier to shoot out through the guides, but accuracy may suffer.

Rear taper

Located at rear end of the belly, the rear taper provides a transition from the large diameter belly to the small diameter running line. A longer rear taper makes the cast smoother because the transition is less abrupt and offers more line control. A shorter rear taper is quicker to cast because the head reaches the guides faster.

Running line

The running line is the longest section in a weight forward line. Small in diameter and light in weight, energy from the rod travels through the running line and loads the thicker head. The loaded head pulls the running line through the guides. A long cast is possible because the thinner line creates less friction traveling through the guides.

LINE WEIGHTS

One fly line is not efficient under all fishing conditions. You will need a line sufficient to cast the largest flies you often use. A heavier weight outfit is a good place to start for those not already settled with what is best for their style of fishing.

A 2- to 3-weight line is ultra light tackle for the most delicate presentations with near invisible flies. But, there are limits to your options when fishing. Casting against the wind becomes more than a challenge; the line is not heavy and may travel through the wind. Distance can be limited if your casting skills are not polished. Another consideration is cost. A rod will cost more because these weights are not standard. You will have to buy a high-end model.

A 4- to 6-weight line can be used for dry flies, small to medium poppers, nymphs, and streamers. This range is general purpose for bluegill. Tackle in the 4 to 6 range should handle most fishing conditions you will encounter. Choose a 4-weight if you want standard rod in the lightest weight. A 5-weight is popular with trout anglers, perhaps the most popular choice for them and an in-the-middle choice for bluegill. Go for the 6-weight if you want one outfit for multiple species, especially bluegill, but also bass. You will pick up bass when fishing for bluegill and then want to tie on a larger fly to catch larger fish.

A 7-weight is heavy tackle for bluegill; it has more capacity to move weighted flies and sinking lines in medium to large lakes.

An 8-weight is more suited to largemouth bass fishing. But, if you want a second outfit dedicated to trophy bass, take a 4 and an 8 and you will have effective tackle for both species. A 4-weight should be sufficient for most bluegill fishing. When heavier tackle is needed for bass or even extreme bluegill presentations, the 8-weight will keep you on the water.

LINE DESIGNS

With so many designs available on the market, the question of which line to use is best answered by how you want the line to perform. Getting maximum performance from a line is vital when fly-fishing.

Buy the highest quality line you can afford. A line lasts longer if you protect it from unnecessary exposure to heat and direct sunlight. A line should also be washed with mild soap when dirty. If you have fished in a lake with foam, you may have noticed a film of dirt on the line at the end of the day.

An easy way to wash the line is to remove the spool from the reel. While holding the spool by the handle, tilt the spool downward toward a basin of soapy water. The line peels off and falls in the basin or on the floor if you want. The beauty of turning the spool over is the line pulls itself off and stops at the backing, like magic.

Weight forward floating

One of the most popular among freshwater anglers, a weight forward floating line is designed for shallow presentations. It will serve the needs of most anglers fishing in ponds and small lakes. For bluegill, the weight forward floating is a good first choice.

As the name implies, the weight is located at the front of the line, behind the tip. A weight forward head is fixed to a running line. A running is a thin diameter line about half the diameter of the belly.

Because the weight of the line is contained in a short head, the belly must be large in diameter. A larger diameter creates more air resistance. Thus, a weight forward line is not the best choice for long distance casting. A short head is pulled close to the guides to cast, but less false casting is needed to load the rod.

A typical weight forward floating line (4-weight) has dimensions such as these:

Tip	0.5'
Front taper	6.0'
Belly	25.0'
Rear taper	5.0'
Head	36.0'
Running line	54.0'
Total length	90.0'

Sinking tip

A sinking tip is ideal when you fish the edges of a plant bed. A line with a floating belly but sinking tip is ideal for subsurface presentation because the approach does not depend on stripping. Stripping is needed at times to control the approach, but too much stripping will defeat the purpose by pulling the fly up in the water column.

A sinking tip with a slow-sinking rate is recommended. The exact sinking rate needed varies with the depth and your patience; 2.0 to 4.0 inches per second is average. A faster sinking rate may appeal to some, but overall a slow presentation catches more bluegill. The key is waiting long enough for the fly to sink deep where the bluegill are taking positions.

A typical sinking tip line (4-weight) has dimensions such as these:

Tip	0.5'
Front taper	2.5'
Sink belly	2.8'
Transition	1.2'
Float belly	21.0'
Rear taper	2.0'
Running line	60.5'
Head	29.5'
Total length	90.0'

Shooting head

Often called shooting tapers, shooting heads are designed for long distance casting. Similar to the weight forward line, a shooting head is also attached to a thinner line called a shooting line. Most of the weight is carried in the head. Shooting heads can be tricky to cast until you get used to throwing one.

As the head launches from the rod, the weight pulls the shooting line behind. Shooting heads are interchangeable. Any number of heads can be looped to the shooting line.

Shooting heads are often the choice of the serious angler—many make their own set of shooting heads in numerous lengths and various sink rates. Shooting heads are popular among steelhead anglers who know how important is it is to get a fly down to the migrating trout.

A typical shooting head (6-weight) has dimensions such as these:

Tip	0.5'
Front taper	5.0'
Belly	22.0'
Rear taper	3.0'
Tip	0.5'
Total length	30.0'

Multiple tip system

If you do not want to carry extra spools to change lines, a multiple tip system is a versatile option in a fly line. Multiple tip systems are available from several manufacturers. An angler can carry a system of interchangeable tips for more options with one line.

Sinking tips enable accurate presentations to pockets in the weeds. For this and other restricted applications, you may prefer sinking tip lines to shooting heads. You may find sinking tips easier to cast than shooting heads, with improved accuracy.

A multiple tip line consists of a 4-tip interchangeable system designed to equip the angler for varied depth and changing conditions. If you do not like connections loops in your line, this is not the line for you. The loops are large and stiff and ungraceful moving through the guides and tip top.

A multiple tip system (6-weight) has dimensions such as these:

Tip	0.5'
Front taper	6.0'
Tip belly	9.0'
Float belly	20.0'
Rear taper	5.0'
Running line	50.0'
Head	40.0'
Total length	90.0'

Sinking line

A full sinking line is easy to cast because the basic shape has a consistent density. Unlike other designs, there are no floating and sinking sections with differing densities to interrupt energy transfer through the line. Inefficient energy transfers can cause hinging, when the line folds at the density transition.

Weight forward sinking lines are designed to sink at a tip-down angle. This angle of descent is possible through an increasing density toward the front taper to reduce sagging toward the fly. Less sag in the line is ideal for deep water. A sinking line is the most effective means of reaching fish in deep water and feeling the strike. Less sag in the line makes feeling a strike easier.

A typical sinking line (5-weight) has dimensions such as these:

Tip	0.5'
Front taper	3.5'
Front sink belly	7.0'

Rear sink belly	27.0'
Rear taper	2.0'
Running line	50.5'
Head length	39.5'
Total length	90.0'

Double taper

Having the same taper at each end, a double taper offers line control at a greater distance because the thick belly runs the length of line. The belly diameter does not have to be as large to load the rod so air resistance is less when the line is cast. Double tapers are popular with some anglers because the tapers at each end are identical and the line can be turned around on a reel when one end wears out. It is like getting twice the use out of a single line.

Anglers who want a smooth rollover for a delicate presentation favor double taper lines. Mending the line at longer distances is also possible with a double taper. But, line design has become so sophisticated that weight forward lines with long-front tapers can be as delicate as a double taper. Double taper lines are good choices for 3-weight and lighter outfits. This may be your first choice for bluegill fishing, instead of a heavy-headed weight forward line.

A typical double taper line (4-weight) has dimensions such as these:

Tip	0.5'
Front taper	6.0'
Belly	78.0'
Rear taper	6.0'
Tip	0.5'
Total length	90.0'

LEADER

The leader is a link between the line and the fly; it can enhance or inhibit the presentation. The length, diameter, and stiffness of material in a leader influence the way a fly sinks toward the bottom. How much control is up to you, but more control helps as the fly sinks.

A 7½- to 9-foot leader is a standard length for a floating line. Tackle companies often list leaders by species to help anglers get what they need. A typical panfish/trout leader is designed with a powerful butt section and a 3- to 10-pound test tippet. Leaders are not often specified for panfish. Scientific Anglers is one company that recognizes panfish as an important species sought with fly-fishing tackle.

Adding a strike indicator to the leader is a personal choice and a mute point if the entire leader works below the surface. You can become quite adept at detecting strikes by holding the line in your hand and training your eyes on the end of the fly line where it bends down into the depths.

Knotless leaders from the factory are a popular choice for bluegill. But, you can make your own leaders to get specific results. Homemade leaders are often tied for sinking tip lines because a shorter than standard leader is desired. A short leader bends less, providing a more direct connection to the fly at the end of the line.

Leaders made for dry flies are also effective and economical. A disadvantage of homemade leaders, wet or dry, is the knots will snag algae. And, if you are in bluegill habitat, algae surround you. Clipping the ends as close as you dare will reduce the amount of algae snagging on a knot.

If you tie your own leaders, start with a butt section stout enough and long enough to turn over the most wind resistant or heaviest fly you may use on a given trip. The butt section should be at least 50 percent of the leader. Recycling a factory leader makes sense because the butt section does not receive a proportional amount of damage compared to the tippet and taper (transition section).

To recycle a leader, add lengths of leader and tippet material to bring the leader back to the original length. When recycling a leader and making one from scratch, do not worry so much about connecting several pieces of material following a gradual decrease in diameter with each piece. A complete leader-making kit is not necessary for bluegill, although nice to have.

If you do not own one, you can buy individual spools of leader material, but premium monofilament line will work if you have a ready supply of line in a range of weights. When connecting pieces of leader material or monofilament line, restrict the changes in diameter to about 0.005 inches or roughly half the pound test for monofilament. The smaller the line, the more difficult this transition step in diameter will become. In other words, connecting 25-pound test monofilament (0.020 inches) to 12-pound test (0.015) is easy, but then going from 12-pound test (0.015) to 6-pound test (0.010) is possible, but not as smooth.

Since leader material is harder and thinner than monofilament line, do not mix leader material with monofilament when building or recycling a leader. And, use the same monofilament brand for all the sections of a leader.

A sinking leader need only be 3 to 5 feet in length. You can cut the butt end and add to the taper to shorten the length of an existing leader. If you are building the leader, you can start with the tippet or the butt section.

Until you get a feel for the process, start with the tippet. For example, say you want a sinking leader with a 6-pound tippet. Cut off 18 inches of 6-pound tippet material (in this example, monofilament). Connect the tippet to 18 inches of 12-pound test monofilament. Roughly 6 inches of line will be lost from each piece tying the knot; both sections are approximately 12 inches. Then, connect 30 inches of 20-pound test monofilament to create the butt section. Tie a surgeon's loop in the other end of the butt and the resulting leader should measure roughly 36 inches long.

The above lengths need not be exact; make each leader a length to suit your needs. You will learn the lengths you need with experience. One important length to remember is the butt section, which should be about 50 percent of the total length of the leader.

Fluorocarbon is fast becoming the material of choice for leaders and tippets. Fluorocarbon is worth the price in crystal clear water, but not all fluorocarbon brands are equal in quality. If you find a brand you like, by all means take advantage of the technology.

When I modify a leader in the field, I at times use a surgeon's knot because the knot is easier to tie standing in the water. A surgeon's knot is a good choice when joining two pieces of monofilament having considerable differences in diameter (a quick fix). A surgeon's knot is stronger than a blood knot, but the blood knot is more streamline.

LOOP-TO-LOOP CONNECTION

To speed the process of changing leaders, make a loop in the end of your fly lines, sinking tips, and shooting heads. A loop at the end of the fly line enables a loop-to-loop connection. Whip finishing the loop makes the connection smooth and strong.

But, if the line is made of monofilament, a whipped loop is going to slip under pressure. Monofilament is sometimes used for slow sinking tips (such as the Quad Tip by Scientific Anglers) and solid-core shooting lines. The answer is a braided loop. Orvis sells high-quality braided loops.

If attaching a braided loop to a monofilament line, add a drop or two of glue to the end of the monofilament inside the braid and another bit of glue around the loose braid at the other end.

A loop at the end of the leader is also a valuable time saver. A surgeon's loop is a large knot, but it is strong and it holds. Make sure the loop is constructed with a double overhand knot.

Tie a loop using a pair of pliers to pull the tag end tight. The secret to an effective knot is tightness. Lefty Kreh, the fly-fishing legend, says a knot fails because it was not cinched tight. This also makes a smaller knot. Make sure you have strong tools in the field to cinch the knots in the butt of a leader. Spit on the knot to lubricate the line before pulling tight.

Use a surgeon's loop when tying a fly to the tippet; it is not necessary, but a loop creates a more effective presentation. A loop gives the fly more freedom of movement. A large loop can be teased down to a small loop if it is not cinched tight beforehand.

BACKING

Backing is not as important while fishing for bluegill, but backing is everything when you need it for a largemouth bass.

Gel-spun backing, made from braided polyethylene, is much smaller in diameter than braided Dacron. Gel-spun backing is abrasion resistant and can withstand direct sunlight and exposure to chemicals.

Dacron will serve the needs of bluegill fishing unless you are using a large-arbor reel. Large-arbor reels have less room on the spool for backing. If you are buying a large-arbor reel and are not sure of the spool capacity, gel-spun backing is a safe bet.

Use an arbor knot to attach Dacron backing to the reel and make sure the knot is tied tight, no matter what type of backing you are using. Test the knot by pulling before reeling the backing on the spool. Scientific Anglers recommends a bimini twist to connect gel-spun backing to the fly line.

REEL

A reel can enhance or hinder the fishing experience. Like using an underweight line to cast a heavy fly, a cheap reel takes the fun out. Any reel will do to store line. And, palming the reel is an honorable way to fight a fish. But, before you buy your next reel, decide whether you want it to be a storage device or a part of the machine.

Large arbor reels give definite advantage to the angler, such as no more frantic cranking to avoid losing another fish from slack. With slack in the line the fish has more freedom to shake its head back and forth as it fights—more opportunity to toss the hook. One good shake is all it takes. Winding a traditional reel as fast as possible may not be enough to remove the slack before it is too late.

Many anglers have switched to large arbor reels because they do not like the line coiled in narrow loops. You may also like these reels because a turn of the handle retrieves more line. If you like to reel a fish in and not strip it, then a large arbor is for you.

A reel suitable for bluegill does not have to be expensive to perform well. A reel of about the same cost as your rod is a reasonable assessment for tackle. Set the reel to wind with your left hand if right-handed. Not switching hands to crank the reel is another way to reduce slack.

A reel—or a rod for that matter—is like a car. Any car will get you to where you want to go. You just have to decide to have as much fun you want to have getting there.

Many reels are machined from bar stock aluminum with close tolerances, features designed to last a lifetime. A quality drag is one of the most important features to seek in a fly reel. A smooth drag is one that does not have to overcome high inertia to start working. Inertia is the tendency of the spool to remain at rest—often called start-up friction.

Drag adjustment is another factor to consider when choosing a reel. Is the drag adjustment knob located where it is easy to find without looking down at the reel?

ROD

A blank made from high modulus graphite with a medium-fast action (mid-flexing) makes a fine rod for bluegill fishing. The rod is more forgiving of casting strokes. Consistency is the secret to graceful and effective casting. Consistent speed and angles in the forward and backward strokes does it. Knowing when to stop the backward stroke and then begin and accelerate the forward stroke is so simple, but so misunderstood, and far easier said than done.

If you are an expert caster you may prefer to use a fast action (tip-flexing) rod. A fast action is designed for dry flies. As with any other tackle, more technical tackle is more expensive.

Length

A 7'6" to 9' rod should serve most needs in an average bluegill lake. A 9' rod will give more line control and increase casting distance, but will be difficult to wield in close quarters with trees and bushes all around. Most lakes and ponds are surrounded by trees, and a shorter rod may be a better choice if you fish from shore most of the time. And, 7'6" is a standard length for some brands of 4-weight rods.

Panfish rod

Rods are advertised by length and line weight, although some companies also list their rods by the species sought. Many companies sell rods for panfish, but a trout rod may provide greater utility when needing to cast flies long distances and price is not a concern. As with any area of fishing equipment, choices are wide ranging. Anglers enjoy near unlimited options in quality fly-fishing tackle suitable for most game species.

Trip rod

A rod should enhance your mode of fishing. A boat trip is not the same as a shore or wading trip. If want to use a fast-loading rod to fish from a boat, overload the rod by one-line weight, such as using a 5-weight line on a 4-weight rod. Some rods can be overloaded two line weights and still perform well.

The benefit of a faster loading rod is short-line performance when you are in close. Short-line loading helps when tucking a fly under tree limbs on the backward and forward cast or pocket casting to open water in a plant bed, to get the line out in tight places where a roll cast is not always practical.

An angler on shore or wading has different needs than the angler floating in a boat. You may at times want tackle that can cast extreme distances. Casting distance often becomes a limiting factor if you cannot wade close enough to prime habitat, which is not always accessible from shore.

Rod length is important if you want to make a long cast. The farther you can cast, the more water you can cover from a single location. You will spook fewer fish stumbling around near shore. A 9' rod is sufficient for your long distance needs.

Going for bluegill with a dry fly is fly-fishing at its best, the way it was meant to be. You can fill a basket with delicious meat. The fish are plentiful, colorful, powerful, and more than willing to take a fly.

A healthy pond has at least one huge bass in it. If you want to switch to a larger fly when the whale appears, you will need an outfit heavier than average for bluegill. Your other option is a light rod for bluegill and a heavy rod for said whale. Choices in equipment should begin with the line, one sufficient in weight to cast the fly.

6

FLY-FISHING

Bluegill are a dream come true for people who like to fly-fish. Many of us do not live near pristine trout streams, but we enjoy fishing for bluegill and largemouth bass. Some of us are simply tired of dealing with high numbers of anglers and low numbers of trout, and others are ready for new challenges in new waters.

MATCHING THE HATCH

Information on fly-fishing is available as never before. A graduate level of scientific knowledge can be disseminated in a book or a series of magazine articles. Matching the hatch should come second to mimicking the behavior of an average insect. Most anglers can match a hatch, too well at times. Matching the hatch follows the law of diminishing returns. There is a point after which the pattern loses effectiveness because the realistic design interferes with the behavior.

Matching the hatch means nothing—to a bluegill or a trout for that matter—if the fly behaves in a strange manner or looks out of place. The image created by a fly is more important than how close the fly matches the hatch. It is better to focus on making a fly behave in a credible manner rather than worrying about a fly looking exactly like the real thing.

FLY BEHAVIOR

Some anglers may argue that behavior is nothing more than presentation. Presentation is getting a fly to the fish. Behavior is what a fish sees when the fly appears. A fly has to behave in a manner that makes a fish recognize food.

A silhouette not so exact but acting naturally fools more fish than an exact match acting strangely. To act naturally means the fly moves with recognizable motion or displays a deceitful image.

It is not difficult for a presentation to look strange; most presentations look strange or we would always catch fish. The fish are there most of the time, but the strikes are not.

If the fly looks stiff, the tippet may be too thick for mass of the fly or the fly was not attached with a loop. A stiff fly may look like another piece of debris not worth the energy needed to strike it. The reasons for a failed presentation are too numerous to list. Focus on what causes a strike rather than what prevents one.

A fly pulled too fast is not natural, but often triggers a strike when presentations with gentle tugs do not. The fly coming into view can be more compelling if it acts in a familiar manner or at times a manner completely unknown to the fish. What is important to remember is the habitat being fished, such as plants or open water, and the amount of pressure put on the water, either public water receiving constant pressure or private water receiving little or no attention from anglers.

Fly-fishing is the same as bait fishing in that you cannot just lay a fly on the water and expect favorable results. You have to fish in the habitat holding potential to attract fish. Then it is within the habitat that specific targets should be chosen. Pick the same targets where an angler would ply bait: plant beds, brush, rocks, points, mounds, and docks. Each cast should be aimed at the edges of habitat such as a plant bed; the holes in the plant bed; the water over the plant bed. The same general points that should be presented to if the habitat were brush, rock piles, and whatever form of habitat is available. Make each cast with purpose; the intent is to present the fly with accuracy, in the narrow zone of maximum potential.

Changing to stay the same

An effective presentation may only be effective under certain conditions. Changes are needed to modify a presentation that works. These changes may include modifying the leader length, changing the tippet diameter, using a smaller or larger fly, using a color more suited to water clarity (dark in dirty water, light in clear water) or one more compatible with the color of the lake bed (more green near plants or orange in open water).

Presenting a fly in a realistic manner is easier said than done. If you are a deer hunter, the importance of how a fly appears is easy to imagine. How many times as you walked through the woods did you think for an instant that a distant log was a bedded deer? As you moved closer or looked through binoculars, the log looked nothing like a deer. But at the first glimpse of the silhouette, the angle, the position on the hill, you swore it was a deer. You had been looking for deer and anything resembling one caught your eye.

Bluegill are no different than any other hunter/predator. They are looking for live animals, too. For a fly to be effective, it has to look like it belongs—like a deer bedded on the hill. Deer bed just below the top of a hill; the position on the hill is important. If you see something looking like deer but out of place, like in the middle of a cut field, being tricked into thinking it is a deer is less likely.

A bluegill is no different than any other hunter, and may strike for other reasons, but not often enough to make the first impression less important. The color and shape of a fly are important, but not strict. You can get by with an assortment of colors and patterns, mere recognition is all you need to trigger a strike.

Presentation varies, some days strict, other days anything goes. The prey living in most bluegill waters do not move in human distances. Movements by zooplankton and aquatic insects are subtle; most anglers would not see the movements without a microscope. So remember, presentations need to start with tugs, not pulls.

First impressions

Stripping is not a natural way to present a nymph. If stripping is necessary to take up slack in the line, be careful how you strip a fly when using insect patterns. Although stripping has its definite place with fly patterns, stripping an insect or invertebrate pattern is no way to create a realistic image. The key, though, is what works. If a presentation takes gentle tugging or all-out strips to draw a strike, by any means necessary, use what works.

Insect larvae move in mere fractions of an inch, if that much. Large insects such as dragonflies and damselflies and some beetles are exceptions. Many invertebrates such as plankton—important to the bluegill diet—suspend in the water column while drifting about. These are too small to imitate with flies.

Other more traditional fly-fishing insects such as mayflies rise to the surface when emerging, and flex back and forth wildly to reach the bottom or vegetation when disturbed, but stripping imitates none of those behaviors. Stillness comes closer to the real thing, but if an odd looking fly sits still in the water, the presentation may not convey any meaning toward food and stripping may win out as the more effective presentation.

False impressions are difficult to pull off when fly-fishing. A bluegill spends much of its life resting and waiting for its next meal. The oldest and often the largest fish have plenty of experience recognizing prey. This experience lends well to their survival in pressured waters. They develop a keen sense of how to avoid hooks without knowing more than what looks good enough to eat and what does not.

Conserving energy is the way to survive. A fish does not know when the next meal will come. While resting, a fish has little else to do but watch the world pass by, and chase the occasional intruder interested in the same space. Watching prey, discerning the good, the bad, and the ugly becomes a matter of necessity. Fish do not see a meal in their world with our point of view. They see countless examples of what to eat and what not to eat, and we can only guess at how they recognize something as food. Perhaps inhaling and spitting out inedible objects often enough makes avoiding a poorly presented fly easy. Presentation of the fly is more important than the fly itself. Presentation is another name for deception.

Refining presentation

You will have days when your best efforts do not bring favorable results. If the pursuit were easy, perhaps anglers would not be so intrigued by fishing with flies.

Unfavorable results often follow sudden changes in the weather. When conditions change, so do the fish. Fluctuating conditions in a lake or pond can make a presentation ideal on one day and worthless on the next. Conditions fluctuate rapidly with the weather and gradually with the climate. The smaller the body of water, the more rapid the changes take place, both good and bad.

Refining presentation makes sense when conditions change. When the conditions change, the activity level of both predator and prey change as well.

Angling pressure on a population of fish increases their ability to avoid getting hooked. Pressure makes fishing difficult. An adult bluegill becomes conditioned to the disturbances caused by anglers. The survivors grow increasingly wary as they notice that not everything that looks like food acts like food.

Popular waters become an acid test for the angler presenting flies. Each time a fish is caught, the experience is imprinted as deceit, in particular when it is fooled by the image of a fly not accompanied by sound and smell. In a single season, a bluegill may be exposed to enough anglers that just the depth at which a fly enters the fish's vision is enough to ruin the image.

If the strike is immediate, the presentation is working as well as you need it to. If a fish moves toward a fly but does not strike, the image presented at a glance looked good, but on closer inspection, failed. Most often this occurs out of sight of the angler. A presentation short of irresistible draws a no-intent closer look at best.

If you can see the fish, notice the position before it moved toward the fly. Did the fish have to move up or down in the water column to get a closer look? If you can see the fish, continue using the same pattern in a different color or a similar pattern in the same color—try to drift the fly directly to the fish.

A fly appearing within striking distance does not pass without notice. Instinct prevails in nature. The longer a fish lives, the more experience it gains toward conserving energy. This sharpens the ability of a fish to recognize a fake. A fish knows well enough not to miss an easy meal, and to avoid a costly one. Most fish surviving a single season have learned to survive through avoiding predators and storing energy. Those fish surviving long enough to reproduce behave in ways that conserve energy through feeding efficiency. Thousands of ancestors in the gene pool have proved this is the best way to survive in the harsh environment we call nature.

Presenting a fly with success requires much attention to detail. An effective presentation makes a fish react. A fish does not follow a complex thought process to strike; it cannot think. A strike is an impulse stimulated by visual recognition. Positive recognition pulls the trigger. Negative recognition produces a wait-and-see response. And, for each instant a bluegill hesitates to strike, the chance of refusal increases.

Live animals move with an unmistakable fluid motion. Their movements create sounds and pressure waves that predators can detect. Predators also detect scent. Most fly patterns used today present a visual image without these other stimuli. Therefore, the presentation has to be exceptional to fool the wise old fish. The common belief that bluegill will bite anything is nonsense.

Proper behavior

Matching the hatch is a good start. Proper behavior is the final step for an effective presentation. Cast each fly with the purpose of hitting specific habitat. And once a cast is made, try to make sure the behavior fits the image a fish has in mind. Instinct will take care of the rest.

Dry flies resemble something on the surface, a silhouette not the same size as the insect imitated. The actual size is not fully depicted on top, which is why you can get away with flies larger than what bluegill normally eat. A dry fly to be most effective has to float high on the surface tension. A semi dry fly floating at the surface is not as effective because the fish has a better view of the offering and its behavior.

Nymphs and wet flies should be tiny to depict underwater prey. Insect larvae and zooplankton eaten by bluegill are difficult to impossible to imitate on a size-for-size basis. Streamers are effective in small sizes. If you are trophy hunting for the few bluegill large enough to strike a minnow-sized streamer, size is not such an issue, but hooking the fish is another concern. Bluegill are not as easy to catch on long-shanked hooks used to tie streamers.

What is the proper behavior? An image that appears real the instant a fish makes eye contact. A fly drifting along in the surface film or slowly sinking to the bottom gets undivided attention, whereas an unidentified object racing through the water column may spook a fish if it is not starving.

A gentle wind is a friend. Wind pushes the water and will move a fly as well. Wind-induced movement either on the surface or below is the most natural presentation you can hope for, if you have the patience. The fly barely moves and fishing is slow, but effective.

The wind is the greatest benefit for concentrating plankton and aquatic insects. The bluegill will move to windswept shores to feed. A strong wind coming to shore will make casting from shore difficult with light tackle.

CLOSE QUARTERS

Make short casts when entering new water. Bluegill can be spread out in front of you without your knowing it. Casting a fly line over them will spook the adult fish. They may not strike when your fly gets to them. Make successive casts in a semi circle (or circle if you can) to reach the most water from a single location.

Even though the bluegill can be anywhere, be prepared by casting to specific targets such as plants or wood. Be confident you will find fish; your confidence will reward you. As you cover the water closest to you, move on or make casts longer. Always make each cast with a deliberate intent to catch a fish. Never let your attention wander, thinking there are no fish and you are here to have a good time anyway.

DRY FLIES

Flies that float should be fished with as little movement as possible. At the beginning of the presentation, let the fly settle for at least 15 seconds. If no fish strikes the fly, give it a twitch, just enough to disturb the surface.

A gentle twitch also makes the feathers or materials of the fly twitch. Most bluegill cannot take much of a twitching movement. But, you have to be where the bluegill are to be successful.

A series of twitches is what catches the fish, but twitching is not always necessary or the most effective means to present a fly. Still presentations, with no twitching, work as well if not better during active times when you are on top of the fish. At other times, stripping attracts more strikes. Stripping should be part of the approach when still presentations and twitches are not effective. If you know you are where the fish are, keep changing your presentation until you find what works.

WET FLIES

Presentations of wet flies and streamers should follow the same priorities as dry flies. Strive for the lightest splashdown as possible and aim for specific targets. A light splashdown is important for wet flies, too. A fly that lands with a forceful splash often repels the trophy fish if they are in a neutral or negative feeding mood, which is much of the time.

Wet flies can be still-fished. You will experience times when no movement is necessary, just sinking without any other movement. This free-drop presentation works when the fish are active or inactive; just let the fly sink—give it a twitch now and then.

You will also experience slow times when no matter what time of day, how good the fly, how active the fish, still fishing doesn't work. You will need to change the presentation by adding movement. Strips should be a hand width at a time. Start high in the water and gradually fish deeper with each cast until you find the strike depth. Waiting longer before the retrieve increases presentation depth. Fly-fishing has no set rules. Strip slower if faster fails.

Streamers are designed to be stripped. When bluegill are actively feeding, streamers work well to imitate small fish, dragonfly nymphs, damsel flies, and juvenile crayfish. Stripping may coax the fish into a reflex strike, no matter what the feeding mood. Just remember, your shortest strip is probably far longer than the real movement made by the prey. Slow and easy is the way of the streamer. Use just enough movement to make the streamer travel. Too much movement will bring the streamer to the surface; too little will let the streamer sink to the bottom.

PATTERNS

Flies should be your personal favorites. You will fish with confidence when using proven patterns. Presentation is everything and fly pattern is secondary if you follow the guidelines.

What you need in a bluegill fly

>Small, size 12-16
>Mosquito like (dry flies)
>Buoyant (dry flies that float on the water)
>Tied sparse
>Leggy (foam spiders and poppers)
>Nymphs and streamers, weighted with a bead or eyes

Classic dry fly patterns

Any trout fly will work because most are small (size 12-16) and sparsely tied. Here are patterns that will work under most conditions. Color is not so much an issue with classic dry flies as it is with spiders and some wet flies. The best trait a dry fly can have is to float high in the water.

>Mosquito
>Adams
>Blue Wing Olive
>March Brown
>Light Cahill
>Light Hendrickson
>Elk Hair Caddis
>Quill Gordon
>Black Gnat
>Griffith's Gnat

Terrestrial and other flies

>Ant (brown if you tie your own or black if you don't)
>Spider (orange with white legs)
>Hopper
>Cricket
>Popper (white or yellow with legs)

Wet flies

Beadhead Prince Nymph
Beadhead Pheasant Tail
San Juan Worm
Beadhead Crystal Bugger (size 10)

GENERIC PATTERNS

Here are general patterns for those who respect function over form.

Dry fly

Sparse dressing that floats well is the goal. Add floatant if necessary.

Hook: Size 16 Mustad Signature R30
Thread: 6/0 Black
Tail: Wood duck flank
Body: Gray dubbing
Hackle: Brown or grizzly

Spider

Closed cell foam makes the pattern float on the surface.

Hook: Size 12 Mustad Signature R30
Thread: 3/0 Orange
Body: Thin fly foam (2mm), orange
Legs: 3 Rubber strands, white

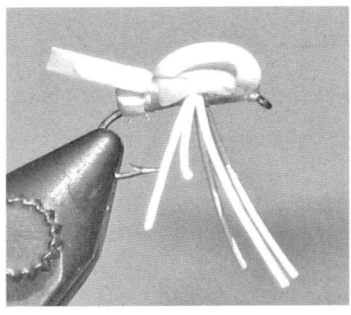

Cut foam in a thin strip and round off both ends. Tie one end in behind legs, first. Then secure in front of legs with a couple of wraps before folding back over to rear of hook. Secure with a couple more wraps and tie off.

Ant

Effective colors vary with the water quality, sunlight intensity, and time of day. Other color combinations to consider: brown/black, black/black, olive/olive. This pattern proves more effective than the foam ant and is as effective in the surface film as it is sinking.

Hook: Size 16 Mustad Signature R30
Thread: 6/0 Black
Body: Antron dubbing, rust
Legs: Ostrich herl, brown

San Juan Worm

Bluegill recognize the San Juan for two reasons. Worms are washed into the water after heavy rainfall and this pattern mimics species of midge larvae. The flesh color imitates a tiny version of a garden worm and red imitates the midge larva commonly called bloodworm.

Hook: Size 12 Mustad Signature R70
Thread: 6/0 Red or tan
Body: Ultra chenille (standard), red or tan

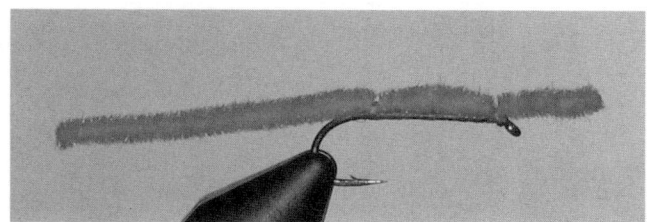

Phostrich

The author's idea of a simple nymph.

Hook: Size 14 Mustad Signature R70
Bead: 1/8-inch* Cyclops, copper
Thread: 6/0 Black
Tail: Pheasant tail fibers
Abdomen: Ostrich herl, white

* 1/8-inch bead recommended by manufacturer for hook sizes 10, 12, 14, but cannot be slipped over barb without pliers. 5/32-inch may be more appropriate if having the barb pushed down is a concern.

Fire Fly

Here is another idea of what I think a nymph should be, provocative. No bluegill can let her pass by without taking a shot.

Hook: Size 10 Mustad Signature R90
Thread: 3/0 Orange
Tail: Marabou tips, orange
Body: Yarn, orange
Rib: Wire, copper
Eyes: Beads, red glass

Attach the beads to the hook using a short length of monofilament line. Heavy monofilament cut from an old leader will do. You want to use monofilament similar in diameter as the hole in the beads.

Cut a small piece of monofilament, about one half of an inch in length. Slide the beads on and melt one end of the monofilament; a touch of a flame is all the heat necessary. The melting line will form a bead, almost an eye itself. Separate the beads with your thumb and forefinger before melting the other end. You want to hold the beads apart the same distance as when attached to the hook. You need to know how much monofilament to melt. The distance between the beads on the eye bar should not be much wider than the hook shank. Tie the eye bar in with figure 8 wraps and glue the wraps before finishing the fly. Tie the eyes in first or just after attaching the tail.

Eyes made from glass beads are not mandatory for this pattern to work, but may make a difference on occasion when conditions are not favorable for fishing or the water receives endless pressure from anglers. If you are looking for that small difference on your fly that can make a big difference in your fishing success, then the eyes are important because no other angler will have flies with red-glass eyes.

And, instead of a tiny piece of a leader butt, you can use colored monofilament to add more contrast to the eyes. Green monofilament is widely available and makes a good contrast with red.

Poppers are one of the most popular styles of flies used for bass and bluegill fishing. Poppers are effective because the buoyancy keeps the fly on the surface. Other dry flies float, but not as well as a popper, which can be made of wood, foam, or plastic. Legs make a popper more attractive. Avoid poppers with long-shanked hooks and long heads. A bluegill cannot get a long popper in its mouth. You will get plenty of strikes, but hooking rates will be low. Commercial poppers are often available in sizes too large for bluegill. As with any other fly used for bluegill, smaller is better.

    A twitch-twitch-twitch retrieve is effective much of the time. Long pauses are necessary at times when the fish are not in a feeding mood. Cast to specific targets during the day. A cast out in the open is effective during cloudy days and in the evening when the light is starting to fade. As the sun sets, bluegill will move up and in, closer to the surface and shore.

7

NEW ANGLERS

Bluegill and other species of sunfish are ideal for children. You cannot find a more compatible species for new anglers. Plentiful and easy to reach from shore, bluegill provide what you need to introduce a new generation to fishing. From catching to cooking, bluegill are it. You cannot ask for more common or better tasting examples.

WHAT'S IMPORTANT

The meaningful aspects of fishing for a young angler are not the same as yours. New anglers lack patience and fishing skills. The most important aspect for most any new angler is to catch a fish, and another, and another... Size does not matter, which may be hard for experienced anglers to accept. Just getting out to go fishing is fine for most adults, but meaningless for the new angler.

The sheer pleasure of getting a bite and reeling in a fish supersedes all else. Any fish is better than no fish. A new angler needs the chance to nurture interest and build confidence from the beginning. The first trip is the most important one, and the need for success from the beginning is paramount to sharing the joy of fishing with a new angler. Do not worry if the first trip is a disaster, especially if you are relatively new yourself. Keep in mind what is important to a child and you will do fine.

Your role

Your role is to keep the children safe, comfortable, and content. Tell them a few basic rules such as no throwing trash on the ground or in the water, and to stay out of the mud. Then, show them how to set the hook after the float goes under. But, only say this once because the youngest anglers will not have the attention span or maturity to comprehend.

Setting the hook is very difficult a young angler to grasp. They will not understand the motion and its purpose. You can alienate them in no time by yelling, "Set the hook!" over and over with no success.

Circle hooks are ideal for bluegill fishing with children. Do not use any other hook but circle hooks with inexperienced anglers. Experienced anglers will enjoy the benefit of using circle as well. Circle hooks are explained later in this chapter.

Constant reminders of what to do and what not to do will steal the joy of fishing from anyone on their first couple of trips. You can ruin a child's interest altogether because those first few trips are the most powerful, regardless of age. Your goal is to provide a safe environment at a quiet pond filled with bluegill. Coaching is not effective until the child is older.

SIGHT FISHING

Another aspect of fishing with children is the thrill of seeing a fish take the bait. Most adults don't recognize the sincere pleasure children get from sight fishing. Don't we all enjoy seeing the fish take? Children are no different. In fact, the absolute smallest fish that can take the bait is just as exciting.

To a child, the 1-inch fish at their feet is more interesting than the trophy possibly hiding just out of view. Dangling bait in plain sight is infinitely more exciting when a fish takes it, no matter what size. To cast to the depths out of sight for even five minutes is a long time for someone who knows not what to expect, but knows just standing there is boring. A child will grow tired in short order. Once a child develops an interest in fishing, the goals for the trip can shift toward the typical goals of the experienced angler.

Do not push the children into trying something new until they are ready. They will let you know when they are ready to move to more challenging tasks. Some will take longer than others to catch on, some will take no time at all and other children will test your patience.

Parents who are experienced anglers cannot appreciate the simplicity of a child having fun doing what looks like nothing important. Too often the parent fails to separate his or her values from what is important to a child.

A child does not have any concept of trophy-sized fish. All they want to do is catch them, often, with little or no effort. And see the action if at all possible. Any dry fly purist will understand this value when fishing for trout. Grown men go nuts over catching tiny trout on dry flies. They spend considerable amounts of money and time so they

Can there be an activity more wholesome and lifelong than fishing? Timeless and fun, what better way can you bring up a child?

can see trout come to the surface to take a dry fly.

Children would rather see tiny bluegill take the bait at their feet than catch much larger fish any other way. Keep in mind the child just learning to fish has no need for a trophy and no patience to learn what it takes to catch one. There is something about watching the fish take the bait that just cannot be fully explained.

SIMPLE SUCCESS

The most enjoyable experience a child can have comes in the shortest period of time with the least amount of effort. Long periods of time between bites can be fatal. When a child has to wait several minutes for each bite, the fun fades. The same can be said of the effort needed to present the bait. If presentation is complex, the fun is long in coming, if at all.

Live bait is best because it is so effective. Live bait has stand-alone appeal; all you do is present it to the fish. For children, this is just what you need for an effective, fun-filled trip. Presentation is as easy as tossing the bait to the water, but the bait has to land in the right place for it to be effective. You cannot catch fish with the right bait in the wrong place.

Another plus of live bait is using a float to present it. Using a float imparts the visual sensation of a fish taking the bait. Most of us remember from our youth watching the float go down. Some of us will never grow too old to enjoy watching a fish take the bait. Who could outgrow a presentation so simple and so effective?

TACKLE FOR CHILDREN

The beauty of fishing with children is their expectations are easy to meet. Some planning and the proper tackle will make a trip a memorable experience. Besides the obvious need for biting fish, methods used to catch them must correspond to the age and experience level of the angler. More specifically, this means bait and tackle.

Cane pole or spincasting rod?

A cane pole is in order for the youngest anglers. A spincasting rod and reel is what most anglers begin with, but children less than 5 years old with no prior experience do not posses the reasoning to crank the reel to see the fish. They can learn to crank the reel handle, but it will take considerable effort and endless patience from both sides.

Another drawback is casting with a rod and reel. A child may not let you cast the bait, insisting that he or she do it, no matter how difficult it is and how much the line tangles. With some children, a rod and reel can be a disaster, and with others, no problem at all.

Parents are the best judge of their sons and daughters, and know who can handle a rod and reel and who would be better off with a

Simple tackle and techniques catch bluegill; you do not have to work hard to have fun with bluegill.

cane pole. If you are not sure your child has the manual dexterity to operate the reel, go by age. Most children 10 years and older should be able to operate a rod and reel; for younger children put a cane pole in their hands. With a cane pole all they have to do is swing the pole over their heads to the water. Swinging a pole comes naturally to a child. The motion adds to the fishing experience.

When a child is ready to move up, a spincasting rod and reel is the next step from a cane pole. Before leaving the cane pole behind, the child should understand what fishing is all about and be coordinated enough to use a rod and reel in the yard. Again, the parents will have to decide where to start (cane pole versus spincasting tackle) and when to move up. Each child will be different in initial skills, attention level, and desire to go fishing.

Fishing with a rod and reel poses a different challenge than using a cane pole. Not only does the rod have to be whipped and then stopped at about the 2 o'clock position, the line has to be held in place then released as the rod is stopped to complete the cast. This is not an easy movement to teach an infant. Reels difficult to use for a child will breed frustration, then anger and boredom if nothing else.

A young child can use spincasting tackle, with practice.

When selecting a casting rod, choose a rod with a pistol grip as pictured on the left; it is easier to hold when casting and retrieving line. A casting rod with straight handle is pictured below.

If you do not know what is best for your child, start with a cane pole. They will not know what they are missing by starting with a cane pole. You can quickly move up to a rod and reel when the time is right. Moving back from a rod and reel to a cane pole may not be so easy.

Tackle should be as simple as possible for children of any age. When they are older and have gained experience, they will want more challenge and the tackle that goes with it. Then is the time to introduce the little ones to new challenges, starting with new tackle.

The experienced child will love to get new tackle. Most cannot wait to use a more demanding outfit. But, don't be surprised if some still like to bring their cane pole along, too. Let them fish the way they want to; if you don't, you will spoil their fun. Letting them have fun, their way, is the surest way to endear someone to fishing, for life. You will have to be patient and not worry about making sure they have fun, just being there is often enough.

The experienced child will love trying new tackle.

Not just any hook will do

More import than the proper rod and reel (or cane pole) is the hook. The wrong hook will bring disaster. Most often the hooks used are too large. Worse than a hook too large is a hook too small. The fish swallows a hook that is too small. A swallowed hook creates a bloody mess and normally kills the fish. As a parent you do not want to deal with swallowed hooks. And your child will not want to see you in despair wondering what to do with a bleeding fish.

The sight of blood coming from the fish is not a pretty sight. An injured fish turns belly up when released. This is the last thing a child should be exposed to. Bleeding and dying fish can have devastating effects on a child's interest in fishing.

Don't be discouraged by such a dismal description of what may sound like a no-win situation. Don't be concerned about what might happen. Instead, prepare for these circumstances and learn how to avoid them. Preparing for a situation means you know what to expect ahead of time, making it much easier to handle the situation.

If you are using a standard hook and it has been swallowed, avoid comments that may indicate something is wrong. Simply cut the line just above the mouth and let the fish go with the hook still in the fish. The hook will gradually dissolve.

You will have to explain what you are doing when cutting the line. A comment saying you didn't want to hurt the fish is convincing and true. Tie on a new hook and act as if nothing happened. You can then continue fishing as soon as a hook is tied and baited. You can have the time of your life when things go well. And, if you think you are having fun, imagine how much it means to your child, to be outside in the fresh air, with you.

Cutting the line and leaving the hook in may save the fish. The chance of this happening is much higher than if the hook was removed. The gills are extremely delicate, blood-filled organs. If you nick the gills in the slightest way while trying to remove the swallowed hook, the blood begins to flow. The wound does not heal; it is a mortal wound.

Setting the hook

The youngest anglers do not learn the importance of setting a hook until they gain considerable experience. A hook not set fast enough is often swallowed. Trying to teach the child to sweep the rod back is easier said than done.

The first thing a child will do is watch the float. If they are really young, they will be mesmerized and either have forgotten what to do next or never have understood in the first place. The longer the delay is in setting the hook, especially with live bait, the greater potential for the fish to swallow the hook. Fast reaction time is critical and too much to expect from a young child. They do not grasp the need to sweep the rod back as the float goes under water. Setting a hook requires maturity as well as hand-eye coordination. Children inevitably wait too long to set a hook.

Circle hooks

A circle hook is the best design to use when taking inexperienced anglers fishing. Not having the hook swallowed is a great advantage when fishing with children. Avoiding swallowed hooks is good for adults, too. A circle hook is recommended for all anglers fishing with live bait. Make every effort to use circle hooks.

Using circle hooks mean carefree fishing; you do not set the hook. Simply reeling in after a fish takes the bait is all that is needed. Circle hooks are more effective when the fish are given time to swim off with the bait, hooking itself, which is how children fish naturally. Either the fish hooks itself or you do.

Using a float

Teaching a child how to catch a fish is much easier when using a float. The child will experience less difficulty learning to set the hook when the float goes down. Setting the hook is a skill not automatically learned by inexperienced anglers, and not needed if using a circle hook. One thing is certain; they will connect the float going down with a fish on the line.

One mistake too many parents make is not using enough weight

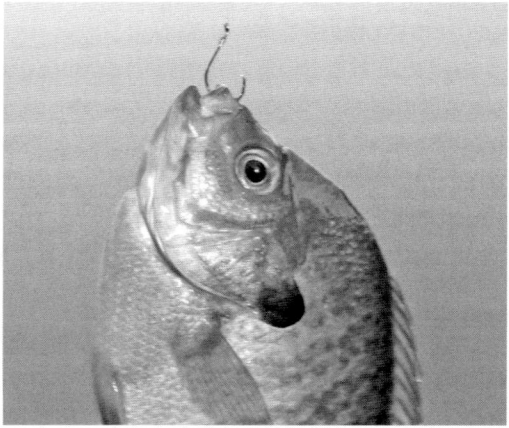

Circle hooks are the most important tackle to bring with you when fishing with children. Circle hooks rarely snag the gut if swallowed, a difficult and unavoidable problem with standard hooks. O. Mustad & Son (USA), Inc. makes circle hooks small enough for bluegill. Use the extra fine wire models, size 6.

to make the float neutrally buoyant. Add as much weight as necessary to make it almost sink. When properly weighted, the float will sit upright with the top color at the water line. Weighted this way the slightest tension from a fish will pull the float under.

Make sure the float is upright in the water. A typical float has two main colors. If the bottom color is under the water, the float should be properly weighted. Tackle does not have to be complicated, but using tackle properly is a big advantage.

Line

Do not buy the cheapest line you can find. If you are not familiar with line, any premium monofilament in 8-pound test is a good choice for most conditions encountered while fishing for bluegill. Lighter weight lines are thinner in diameter and less likely to spook fish, but lighter line breaks more often. Then you will be tying more knots, so there are trade-offs. If knot tying does not bother you, use 6-pound test monofilament. Using lighter line, the line has to be freshened by cutting off the rough parts as needed. The heaviest line you should consider, 10-pound test, is an option if you do not want to deal with broken lines and tying knots. Children will drag the line over every rock and snag every bush in sight, and some you did not know were there.

If you start with a cane pole, tie the line so it is as long as the pole, perhaps a foot longer at most, including the hook. Making the line the same length as the pole will bring a fish to your hand when you raise the pole and swing it back.

If you are purchasing a reasonable spincasting outfit, it will come with premium line. The decision of what line to use will not be an issue, use what comes with the reel. The line should be replaced each year and more often after frequent use around rocks and logs that result in numerous snags.

Fishing trips

The most important thing to remember when taking a child fishing: The trip is for them, not you. There is no opportunity for you to fish. You cannot expect to fish yourself and also be the mentor your child needs you to be. Children need undivided attention at all times if you are going to do it right.

If you have to fish, don't take the children; they will not experience fishing to the fullest and may not ever take a liking to it. Later on, after your children learn to fish, they will want nothing more than to have you fish with them.

Choosing a location

Remember when you choose a place to go, you are taking children fishing. You want an open area free of biting insects and not thick with trees and brush. Ponds are ideal for this type of a setting. Trees do not surround newer ponds. You also want a pond filled with sunfish, stunted bluegill or not. A pond of stunted (small) fish is best because the pond will be filled with them and they will be lined up at the bank with napkins tucked in their gills.

Again, size does not matter, but frequent bites do. The faster the action, the better the experience.

City and state parks are good choices to investigate, if ponds are present, fish are often present as well. Suburbs are another option. Retention ponds are common in housing developments built in the last 15-20 years. Some cities or the developer will stock the ponds.

If parks and suburbs are not available, call or e-mail the state agency in charge of fisheries. Ask for a recommendation; say you are looking for a pond (small lake) with easy shoreline access for children. A pond with high numbers, regardless of sizes is what you need. A pond filled with algae is not desirable. The algae will cling to the hook and line with every cast.

If you are looking for a place to fish, try the suburbs. Most people have no idea what swims in the neighborhood pond, such as this trophy redear sunfish. Newer subdivisions require a retention pond in the development. Often these ponds contain fish and are open to fishing. Newer ponds have fewer trees growing along the banks, as a general rule. Open areas provide excellent places for new anglers to learn how to catch fish.

Time

Time your trips for the most productive parts of the day, sunrise and sunset. Sunset is the most practical because dragging children out before sunrise is a bad idea until the child is older.

If morning is a better time for your schedule, try to be on location before 10:00 A.M. By midmorning the sunlight intensity normally pushes the fish down into deeper water. Reaching them from shore is more difficult, especially if the children want to cast their own lines.

Fishing slows down or stops when the sun is high. At such times, spend more time looking for fish. You may be able to find active fish in a shaded area. Deep areas close to shore are the places to try.

Overcast skies are an exception. Cloudy skies often mean good fishing through the day.

Comfort

Make sure the children are comfortable. This means providing food and drinks, protection from direct sunlight (clothing, hats, etc.), and chairs. Chairs will not often be used when the fishing is fast. Waterless soap or disinfectant wipes should be used before giving food to eat.

Do not stay too long. If the children want to leave after 10 minutes, leave immediately, forget that it took you 3 hours to prepare and get there. If the fish are not biting, move around the pond and try to find some that do. But, don't drag it out thinking you will find them if it takes all day. If you cannot find the fish, or few are biting, ask the children if they want to leave. If they do, leave with the promise of returning soon.

The same can be said if a child wants to leave no matter how well the fish are biting; it is time to leave. One sure way to kill the interest in fishing is to make the first few trips long and miserable. You do not want to give the impression that fishing trips are long and boring and once you go, there is no coming back.

Safety is your greatest obligation. Each child needs to wear a properly fitted U.S. Coast Guard approved Personal Floatation Device (PFD) at all times when on the water. A PFD, polarized glasses, sunscreen, and cool drinks will keep your child safe and comfortable.

Get your children interested in the outdoors; you will find it healthy and educational; it's quality time you cannot buy.

Camping is a great way to spend time with your family. You will find plenty to do, especially if you want to go fishing for bluegill.

If you prepare a child for the cold, you will be amazed how much patience and interest he or she will have for ice fishing. Do not go unless the ice is thicker than necessary.

NEW ANGLERS • 155

Have a backup plan if the fish do not cooperate or the time becomes boring. A few slices of bread are a good idea. Bread can be used as bait or duck food. If your guests want to explore the pond, let them feed the ducks. They will love to.

Children learn quickly. You will be amazed how fast they learn. Soon they will gain the skill and understanding of adults. What parent would not embrace the activity that helps children learn about our natural resources and have a blast doing it? Would you like to see a similar expression on the face of your son or daughter?

No explanation needed.

8

LAKE STRATIFICATION

Stratification occurs when a lake separates into layers. The layers of water are separated by temperature and dissolved oxygen. Once a lake becomes stratified, dissolved oxygen becomes a limiting factor in the bottom layer, essentially making the bottom layer inhospitable due to poor quality of water. Water quality in this sense does not mean pollution. Water quality can deteriorate during the natural cycles of sunlight intensity and decomposition. (Think of a stuffy room without air conditioning when the air is hot and humid.)

Temperature and dissolved oxygen concentration are linked; when one changes, the other changes, too. Of these two conditions, temperature determines if a lake will become stratified. Water will settle in the lake basin according to the temperature of the water.

WHY YOU SHOULD CARE

Stratified layers can restrict portions of a lake where bluegill feel comfortable. With comfort in mind, bluegill choose positions where the water is comfortable and close to a source of food.

Bluegill suspending over deep water is easier to understand if you know what they are eating. The tendency to suspend over deep water can be driven by the desire to eat or to select a more comfortable location (water quality).

Food sources will determine the general locations where bluegill spend time in a lake, even if the stay is temporary. A bluegill will take more specific positions based on water quality, including sunlight penetration. This behavior is not universal; the bluegill living in a lake will not all choose the same location. Bluegill living in one lake will act the same as the bluegill living in another lake. Even though water quality conditions and food sources will not be identical, the basic movements will be similar. (Early in the morning the bluegill will be in shallower water or closer to the surface. As the sun rises, the bluegill will descend to deeper water. The reverse happens at the end of the day.)

DENSITY AND TEMPERATURE

Water is unique. Unlike other compounds on earth, the density of water changes with changes in temperature. Water is most dense at 39 degrees Fahrenheit. Water becomes less dense when the temperature ranges from 38 to 32 degrees. The less dense water in this range rises above the 39-degree water, which is why ice forms on the surface and lakes do not freeze solid. This explains why ice floats. Temperature affects the feeding activity of bluegill. Water too warm can make bluegill quit feeding, just as cold water can.

LAYERS

Mixing is necessary to transport dissolved oxygen and nutrients through a lake. As water warms and cools, it rises and sinks with changes in density.

Mixing in most bodies of water can occur if the wind is powerful enough to push the water from the surface down into the depths. Tributaries pouring in and dams letting water out can also induce mixing, but rarely move enough water to mix an entire lake other than after extreme rainfall.

Layers form if mixing is blocked by a denser layer of colder water.

The dense layer may shift back and forth within the body of water, but for the most part, remains intact until a seasonal change in the weather breaks it up.

Epilimnion

The upper layer of water that forms in a stratified lake is called the epilimnion. The epilimnion is the living layer. Sunlight provides the energy for the food chain. Wind and solar heating mix the epilimnion. As this upper layer mixes, oxygen concentration remains high from constant exposure to the atmosphere and photosynthesis. With plenty of food and dissolved oxygen, the epilimnion is comfortable for the bluegill.

Thermocline

Below the epilimnion lies the thermocline. A thermocline forms in the summer. This layer is more of a temperature gradient, a rapid plunge to cold water. Temperature drops more per foot of depth than in the other layers. If you have ever dove deep in a stratified lake, you will remember hitting the instant cold of the thermocline.

The thermocline is a thin layer compared to the epilimnion. Once established, the thermocline blocks the downward movement of water. Complete mixing in the lake ceases.

Hypolimnion

The bottom layer is called the hypolimnion. This layer is thick, similar in scope to the epilimnion. Unlike the epilimnion however, the hypolimnion lacks the water quality to attract and hold bluegill.

The lack of sunlight penetrating the hypolimnion means no oxygen replaced by photosynthesis. The water here lacks oxygen because the decay of organic matter uses oxygen. No mixing with the upper layer of water (epilimnion) means no source of atmospheric oxygen to the hypolimnion.

A lack of oxygen in the hypolimnion does not cause a problem in every body of water. If the water is not very fertile (oligotrophic) oxygen concentrations do not drop too low to sustain fish. The amount of organic matter decaying on the bottom is not using all available oxygen that is otherwise available to the fish.

Oligotrophic lakes are very clear and mostly located in remote areas away from civilization. Lakes in Canada are often oligotrophic.

Seasonal changes

As the ice melts and water temperature begins to rise in the spring, the water sinks if the temperature is above freezing (32 degrees) and less than 39 degrees. The water mixes for a short period and settles with the warm water at the surface and the denser colder water at the bottom. This period of mixing is called the spring turnover. Dissolved oxygen is distributed throughout the body of water.

A lake stratifies during the late spring and early summer. The long days of intense sunshine warm the surface water until a thermocline of colder water develops where the sunlight does not reach.

The thermocline is the floor for bluegill activity during the summer. Bluegill can swim through the thermocline, but if they do, they won't stay. If you have ever walked into a building filled with egg-laying chickens, it will take your breath away. You won't die, but you will get out as soon as possible to get a breath of fresh air.

Autumn, like spring, is a time of change. As the air temperature decreases, the water temperature at the surface follows, especially at night. As the temperature decreases, the density increases. The water sinks as it cools. Sinking water falls through the thermocline, mixing the lake down to the bottom. This event is called the autumnal turnover.

In the winter, water cools rapidly at the surface until ice forms. A layer of ice prevents wind-induced mixing. Ice prevents super cooling from exposure to air, and the mixing that caused the autumnal turnover ceases. Ice also allows snow to cover the water. A prolonged cover of snow blocks sunlight necessary for photosynthesis, which continues at a reduced rate in the winter. What little photosynthesis occurs is enough to prevent fish kills.

The lake essentially stratifies through the winter until a turnover occurs in the spring. The water is 32 to 38 degrees just below the ice and then 39 degrees down to the bottom and at the greatest depths it is dark and possibly stagnant.

YOUR STRATEGY

The first thing to do when wondering whether the lake is stratified is not to worry. If you are not fishing in a boat, you are not likely to be fishing in the hypolimnion, unless you are standing on a dam. If you are in a boat or fishing from the dam in the summer, don't fish on the bottom.

Sunlight penetration

The key to success in a stratified lake is to keep your bait in the epilimnion—the euphotic zone—the portion of water receiving sunlight sufficient for photosynthesis. If you are in a boat and have a recording thermometer, just find the depth where the thermocline begins. (The temperature drops rapidly in a span of 3 or more feet.) Then, keep your bait above that depth and next to habitat of some sort.

If you do not have a recording thermometer in your boat or are fishing on the bank and have ready access to deep water, you can estimate the depth of the euphotic zone. If the lake contains water typical in many lakes, not exactly clear, the euphotic zone averages 8-12 feet deep. Clear lakes more common in the western states and

Florida will have euphotic zones much deeper than 8-12 feet. Turbid (murky) lakes will have shallower euphotic zones, no matter where the lake is located. Sunlight is scattered by the suspended particles in the water.

The easiest way to stay in the euphotic zone—especially when fishing from shore—is to fish where aquatic plants are growing. The zone where sunlight reaches the bottom of a lake is called the littoral zone.

Ice fishing requires the same strategy; avoid the bottom in the deepest areas of a lake. Mixing ceases with the forming of ice and oxygen depletion occurs in the bottom layer. As in the summer, the processes of organic decay consume dissolved oxygen concentrations and photosynthesis is not sufficient to replenish dissolved oxygen concentrations.

LAKE STRATIFICATION • 165

Ponds do not stratify if the depth is shallow. Sunlight penetrates a large proportion of the water in a small pond and the wind has considerable effect. Mixing occurs throughout the water column.

166 • CATCHING BLUEGILL

A tide clock for bluegill fishing? You do not need to see tidal movements to predict the potential periods of active feeding.

9

LUNAR PERIODS

What on earth does the moon have to do with catching bluegill? Plenty. The position of the moon affects the fish.

GRAVITY

The moon and sun exert a force of gravity on the earth. This gravitational pull creates the tides in the oceans and seas. Does the gravitational pull affect lakes and ponds as well? Yes. Think of the times you waited for an elevator. You watched the numbers change above the door while you waited. Did you ever watch a see-through elevator going between floors? Watching the elevator move is not the same as riding in one; you cannot feel the pull that the people inside feel, can you?

Fish feel the lunar period (elevator ride) in slow motion while we just watch the phases of the moon go up and down. At least this is how you can imagine what the gravitational pull must feel like to the fish.

The moon rotates on an axis as it revolves around the earth. One lunar rotation takes the same amount of time as one revolution around earth, which is why the same side of the moon faces the earth, but the illuminated portion of the moon changes each night. The sun illuminates the portion of the moon we see. The new moon is dark because it appears between the earth and sun, and the side of the moon facing the earth is shadowed.

FOUR PHASES

A lunar period has four phases: new moon, first quarter crescent, full moon, and the third quarter crescent. Beginning with the new moon, the phases begin waxing. This means the illuminated portion (crescent) increases a little more each night.

A waxing moon is easy to identify; the crescent is illuminated more toward the right side. After a week, the right half is illuminated. This is called the first quarter crescent. After two more weeks, the moon is full. A full moon is located on the opposite of the earth from the sun—the side of the moon facing earth is completely illuminated by the sun.

One night after a full moon, the phases begin waning—the illuminated portion decreases. One week after a full moon (three weeks after the new moon), the moon becomes a third quarter crescent—more of the left portion is illuminated. In approximately 29½ days, the lunar cycle ends and the new moon appears. The next lunar cycle begins.

A high tide occurs twice a day, approximately 12 hours and 25 minutes apart, which is about one-half of the daily lunar rotation of 24 hours and 50 minutes. A low tide occurs 6 hours and 12 minutes after the high tide, which is about one-half of the period between high tides.

Water swells from the gravitational pull at two points, the point on earth closest to the moon and the equivalent point on the opposite side of the earth, farthest from the moon. As the earth rotates, these swells follow the moon in its orbit. Tides rise and fall as the earth rotates under this pull.

SOLAR EFFECTS

The sun also affects water by strengthening or weakening the gravitational pull of the moon. Every two weeks—during both the new and full moon phases—the earth, sun, and moon are in line with each other. The gravitational pull of the sun and moon are then combined. Gravitational pull on the earth becomes 1½ times greater than that of the moon alone. The greatest difference in height between a high tide and a low tide occurs during the new moon and again during a full moon. During the first quarter crescent and third quarter crescent phases, the moon and sun are at a 90-degree angle to the earth. The moon and sun become opposing forces working against each other. The gravitational pull is weakest then.

INLAND EFFECTS

The same forces that create an ocean tide also affect freshwater lakes, but to a lesser degree. A lake has a much smaller basin than an ocean. Water in a lake cannot truly rise and fall as in an ocean tide. The water oscillates about an axis near the center of the lake basin. The oscillating movement in a lake is called a seiche.

A seiche in a freshwater lake is not as visible as a rising or falling tide. The seiche takes several hours to make one slosh (oscillation). The change in water level at one end of a lake may only be a couple of inches higher or lower than the level at the other end.

A small example of a seiche occurs when you drive to work with your morning coffee. The sloshing back and forth of the coffee moves in principle as the lake water would. A seiche moves in almost still motion compared to your coffee. The seiche switches ends (rises on one end as it falls on the other, in step with the tides). Just as fish sense changes in barometric pressure, they sense changes in the gravitational pull.

Lakes, rivers, and streams dammed at one end can pile water up at the dam. Wind can also move water. Strong wind can push surface water in the same or opposite direction of normal flow.

DAILY ACTIVITY

Fish are active during the new and full moon phases. Fish are also active on a daily basis during the rotation periods when our position on earth is in line with the moon. This is the freshwater equivalent of tides, when the moon is directly above or below our position on earth.

For landlocked anglers without an ocean, these rotation periods correspond to the solar and lunar tables published in various magazines and booklets. In these tables, major and minor periods predict the best times to fish or hunt on a given day.

Guess what those times are based on? You guessed it, the moon. The best days to fish during a given month are 3 days before and 3 days after the new moon or full moon. Two active periods during each day reflect the hour before and the hour after a high tide. Native peoples have known since long ago that the fish are more active just before and after a high tide. Even if you live inland, away from the ocean, the effects of gravity are still felt.

Predicting the feeding activity of fish in a lake at any given time is guessing at best, unless the angler observes an insect hatch or jumping minnows. Most anglers have their own ideas as to why fish feed at certain times and develop lockjaw the rest of the time. This is not to say go fishing only when the moon and sun align. If you did, you would rarely go.

I have fished many times when the sun and moon were right and came home without catching a single fish. Who knows what happened? One can only guess about other factors such as barometric pressure or wind. I have also enjoyed some incredible fishing when the time was not right.

I am convinced that feeding activity does pick up for a couple of hours each day, during daylight hours when a tide would have rolled through had there been an ocean. On those days when the timing is right, you swear the fish will soon hit the paddles if it gets any better.

Solar and lunar periods are not foolproof guarantees of success. I find these tables most useful when planning trips ahead of time. When planning a trip, you can pick the days months or a year in advance. In between those best times, go fishing every chance you get, as most anglers do; it would be foolish not to go.

A barometer can be a useful tool if you live close to the water and can go fishing on short notice. A moving barometer forecasts a change in the weather. Fish feed actively prior to a weather front. Notice the barometer frame is cut in the shape of a bluegill.

Ice fishing is a great way to pass the coldest months of the year. Notice the parasite (anchor worm) attached to the pectoral fin of the bluegill. The sunken belly of this fish means energy used has not been replenished, which is common during the winter. Wax worms are just as good under the ice as they are in the warmth of summer. Maggots are another option when the fish refuse the wax worms.

10

ICE FISHING

Access is what makes ice fishing unique. You do not need a boat. You can walk out to the fish. If you normally fish from the bank during the warm months of the year, fishing through the ice is an enjoyable change. Fishing in the dead of winter can ease your bout of cabin fever. And, there's no concern about keeping the fish fresh, just put them on ice.

THE WINTER CHALLENGE

Ice fishing is not as easy as one would think. Easy access to the fish does not lead to immediate success through the ice. Although you can walk on the water, finding fish is not so straightforward. Staying warm is another challenge.

Prepare to spend long periods of time on the ice with little movement; you can quickly lose your zest for winter adventure if you become cold and uncomfortable. Plan carefully with full intent to enjoy the challenges of finding fish and withstanding the cold and wind.

Drilling holes with a hand auger will keep you warm, sometimes too warm. Remove your heavy jacket if you get too warm while drilling holes in the ice. It is good to increase your heart rate and blood flow, but not good to get overheated and

Look for bluegill near habitat they use during the summer, the plants and brush they use after spawning. If you do not know where the habitat lies, stay close to the banks, in the vicinity of the dam. Look for water 8-12 feet deep in small lakes, and present bait just off the bottom to about half way up the distance to the hole.

begin sweating, that will cool you down.

Regulate your body temperature by shedding layers; unzipping and unbuttoning are lesser options. You will notice the cold most often when the fish are not biting. When you become cold while wearing all your clothing, it is a good time to move and drill more holes.

Moving and drilling holes will warm you up and increase your chances of finding fish if you have not found them by now.

Finding fish

Unless you want to invest in the latest electronics developed for ice fishing, an effective means of finding fish is to use the favorite haunts of the bluegill, the locations they used in the late summer and autumn. Knowing their favorite haunts in warm weather will help you find fish in the winter. If you are fishing in new water that you have not fished before, then you have to start with the basics of finding bluegill.

The strategy then becomes one of following the contour of the land as it enters the water. The land underwater may be the same as it is on shore before entering the water. In other words, what you see on land will continue with the same slope. A steep bank continuing underwater is what most anglers consider a dropoff. The same can be said of creek channels with rocky ledges, but these areas are not as obvious.

If you are in new water, start with the dam. Unless you are fishing in a natural lake, a dam will be present. The dam represents deep water. The spillway in the dam will show the deepest point of the lake basin.

Finding the deepest portion of the basin is important, and not because you want to fish in the deepest water. The deep-

Jigging bait can be more effective than live bait hanging motionless. Popping a jig upward several inches is a popular method of fishing. A quivering action is often more effective than bouncing the jig several inches at a time. Serious anglers often jig two rods at once. Notice how close to shore this angler is fishing. Steep banks and brush attract bluegill, even in the winter.

est portion of the basin gives reference to where you do want to fish, the shallow habitat adjacent to deep water.

WINTER HABITAT

Ice fishing is much like fishing before the lake turnover. Bluegill will be using nearly the same areas as they did in late summer and autumn. They use the same or similar areas to avoid lake stratification. This means they will not be on the bottom in the deepest portion of the lake.

Vegetation is key in any season

The role of vegetation in the life of a bluegill is important all year long. First and foremost, vegetation is the natural habitat for bluegill. Plant growth is not the only habitat used by bluegill, but one of the most preferred habitat types.

Vegetation is not actively growing in the dead of winter, but plants, even in their dormant phase, still attract bluegill. Some plants, depending on species, will turn brown if layers of snow cover the ice, blocking direct sunlight. The physical structure of the vegetation is still in place. Other species will decay down to what botanists call a vegetative stub, the source of new growth come spring. Bluegill prefer to be near erect plants, but will spend time at the broken down plants. When you know the location of plant beds, start there, regardless of the condition. You will not know the condition unless you have an underwater camera.

Winterkill occurs in ponds when a thick layer of snow blocks direct sunlight for an extended period. Plant decay on the bottom uses oxygen. Photosynthesis no longer replenishes dissolved oxygen concentrations. The layer of ice blocks surface water in contact with the atmosphere, another source of oxygen.

Fish kills occur in the winter because the oxygen drops too low to sustain the fish. Large ponds and lakes have a greater volume of water and fish kills occur less often if the water is deep. Larger bodies of water with depth can separate into layers as in the summer. The fish merely move to water in the upper layer where ample concentrations of oxygen are still present.

If you know where the plant beds grew in the summer, start there. Drill holes on the deep side of the bed. If you do not draw strikes in 15 minutes, change the depth of your bait in one or both holes.

If you are one foot above bottom with one line and jigging the other line at about two-thirds of the depth, move one line to one-half the depth and jig with the other at another depth

that you have not tried or that you tried with a still line. The goal is to present the bait in the lower half of the water column. Start fishing about 1 foot above bottom. If you are not getting bites near the bottom, move higher in the water column until you are about half way up in the water column. If plants are present, then present the bait just above the plants.

After about 30 minutes of working the holes, move perpendicular to shore if you did not get any bites. You can move directly over the bed toward shallower water in the direction of shore or deeper water in the opposite direction. You are looking for changes in the bottom structure, such as holes in the plant bed, drops in elevation (if you do not know where the plants grow), and rocks on the bottom.

If you see wood sticking up above the ice, try there if the water is 6 feet or more deep. The wood can be from a deadfall or living trees and brush. Brush in the water near the dam of a pond is an ideal place to start.

Do not stay in one place for a long time unless you are in a very small pond. Keep moving until you find fish; you may have to drill few holes or you may have to drill many, more than you imagined; it is the nature of the game.

Early ice

Many anglers like to go fishing as soon as ice forms because early ice is considered the best fishing of the season. Make sure ice is safe before going. Ice is not equally thick across a body of water. Tributaries and undercurrents can inhibit ice formation. During the early period bluegill can still be using deep water. Oxygen is still plentiful throughout the lake or pond. Not long after ice closes off the surface, oxygen is no longer as plentiful as it was before the ice formed. Dissolved oxygen concentrations begin to drop near the bottom in deep water.

Bluegill leave the depths in search of higher concentrations of dissolved oxygen. They will also move around during the day for other reasons; some anglers think subtle changes in temperature caused by bright sunlight will encourage some to move.

Late ice

Late in the winter, dissolved oxygen concentration becomes a limiting factor in some lakes and ponds. Depending on the specific character of the body of water, bluegill may move back and forth along a slope as the winter proceeds. The nature of their activity depends on water conditions. Bluegill, like any animal, do not go on joy rides; they move only as far as they have to. Just like people, some are more inclined to roam than others.

Deep faith

If you are not familiar with the water in the lake you are fishing, make the dam your starting point. If you then go to the bays adjacent to the dam, you may find vegetation or brush close to deep water, right where you want to start fishing.

In an average 1-acre pond, you will not have to move back from the dam to the bays, you can begin fishing along the dam. The shape and depth of the pond basin will dictate how far away from the dam to start. The shallower the basin, the closer to the dam you can start. Remember the goal is to find water where sunlight reaches the bottom to sustain plant life. In a pond this is normally at the perimeter.

When fishing in lakes, the best places to fish are normally closer to shore than to the center of the lake. Ponds are much smaller in scale; fish may be found more toward the center, if the ice is fresh.

TACKLE

Ice fishing for bluegill can be as simple as it is in the summer. Bluegill are what you make of them. You can catch them with the latest advances in technology. You can catch them with a simple tackle and a sense of the fish. The choice is yours. Not needing a boat to go anywhere may be a welcome change.

Depth finder

A depth finder is the simplest and perhaps the least expensive tool of ice fishing. Unless you rely on expensive electronics, a depth finder is the best way to determine the depth of water where you are standing; it is also an easy way to set your float to present the bait just off the bottom.

As mentioned earlier, the bottom is the first place to set baits, and then try depths higher in the water column if strikes do not come from the bottom. Bluegill will take positions just above the plants and brush on the bottom. Their exact location depends on how high the habitat rises from the bottom. Remember, by bottom, you do not want the deepest part of the lake or pond. You want the bottom of the littoral zone.

Called a plummet in Europe, a depth finder is nothing

more than a vinyl-coated piece of lead with a spring clip attached at the top. Clip the weight to the shank of a hook and lower it to the bottom. A plummet is an important tool to determine the substance on the bottom, if it is a hard, soft, or weedy bottom.

Rod

An ice rod is a miniature rod; it is the same basic rod; the most striking difference is the size. Most rods used for ice fishing are spinning rods, a couple of feet in length with power varying from super light to medium, depending on the purpose.

Any rod will do, including your warm weather equipment. The longer the rod, the farther back from the hole the angler has to be. Anglers prefer a position perched over the hole in the ice. You will not realize the same sensitivity and feel with a standard rod, but a standard rod is a good start if you want to give ice fishing a try.

High-end ice rods often have strike indicators made of a few inches of spring steel. Anglers often add their own strike indicator to the ends of their rods. The monofilament line passes through the eyelet of the strike indicator and it becomes a sensitive extension of the rod. The slightest stress on the bait will make it move. A neutrally buoyant float will provide the same sensitivity, but not as close to eye level; you have to watch down in the hole rather than at the rod tip.

A skillful angler does not need the highest technology, knowledge will do. Merely watching a float or jigging a lure can catch bluegill. The bait has to be where the bluegill are, of course. You cannot expect the fish to actively searching for food.

Reel

Most reels used for ice fishing are spinning reels sized to balance the miniature rods. Spincasting reels were once used, but not as much as spinning reels today. A spincast reel is easier to dispense line from while wearing heavy gloves. You push a button with your thumb and hold it down instead of cocking the bail and gripping the line with your trigger finger. If you decide to use spincasting reel, or a baitcasting reel for that matter, make sure the rod has a trigger. Casting reels are difficult to use on a straight-handled rod without a trigger.

If you are purchasing a rod or reel for ice fishing, you are better off with a balanced outfit. As with the rod, warm weather reels will work, but may be too much for a diminutive ice rod. Grease in the warm weather reel can be an issue. The lower the temperature, the more stiff the grease will make the reel feel when you turn the handle.

Line

Line, like any other tackle, is a matter of choice. Monofilament designed for ice fishing has specific properties. If you are serious about fishing, ice line is waiting. Braided lines, dark lines, and fluorocarbon are available. If you just want to catch fish, any premium monofilament will do. Try 4-pound monofilament test to start.

Hooks

Many anglers rely on tiny jigs and ice flies instead of plain hooks. Tiny jigs have the advantage of putting weight where it counts, at the hook. This keeps tension on the line all the way to the hook, which keeps slack at a minimum. An angler will feel the slightest resistance when a bluegill takes the bait. If the line were weighted above the hook, as is traditionally done with live bait, jigging would not be as effective, because the length of line below the weight would be free to bounce around out of touch. A bluegill would not be detected until it swam off with the bait. The bait can be picked up and spit out without the angler ever knowing. The chances of hooking a bluegill are higher if the angler detects the take.

Ice jigs are becoming more common than ice flies. Ice jigs are tiny teardrops with colored heads; some have a spinner blade attached. Some anglers successfully use bare jig heads with no color. Color can make a difference on some days more than others. Bright colors such as pink and orange and fluorescents are popular and effective. Weather conditions and snow covering the ice may dictate what color is more attractive one day and not attractive the next time.

Jigs can be specific to ice fishing or not; what you want is a jig that hangs horizontal. A jig with the eye on top hangs horizontal. A jig with the eye out on the end hangs vertical. A bluegill can easily slip a horizontal jig in its mouth, but not so with a vertical jig. Standard jigs come as small as 1/32-ounce, which comes with a size 6 hook. This is suitable for ice fishing.

Some anglers prefer even lighter jigs made for ice fishing. Jigs can be too light for the size of the line used. If a jig is too light, the bait will sink slowly. If the line is stiff from cold temperatures, the line may kink outside of the hole in the ice because the jig is not heavy enough to pull the line down the hole. Sufficient weight will put tension on the line so the angler can keep in touch with the bait.

The correct hook can be any hook if it is small. Size 10 or

size 12 will work well with bait such as a wax worm or maggot. Hooks designed for fly tying come in small sizes and are high quality.

Floats

When choosing floats for ice fishing, keep in mind the cold temperatures. If the air temperature is below freezing, ice forms on the line where water collects. Water collects and freezes in many places, the guides of your rod, and especially the holes in a float if it is an in-line slip float.

The in-line slip float, efficient as it is effective in warm weather, is difficult if not impractical to use when ice freezes in the small hole at the top of the tube. The float freezes in place at the hole and needs to be slid down so friction can heat the tube and free the float. Not all slip floats have a narrow hole through a tube, so avoid narrow holes.

Fixed floats prevent the line from being reeled in. The line has to be pulled in hand over hand, which was the common way of fishing not too many years ago.

Slip floats are recommended. Floats are made specifically for ice fishing in cold temperatures. Ice floats are often made of foam. A bead above the float is necessary to catch the float stop.

You can also modify the plain old and reliable red and white round float to work effectively in the winter. To make a round float into a slip float, push out the bottom loop and twist. The loop (curved wire) is not returned to its receiving hole (sometimes a groove and not an actual hole) for a tight grip on the line. Instead, the wire loop is propped open by not putting it back where it belongs. This wire loop then becomes the connection to the line and does not freeze as easily as the in-line float with the tiny hole at the top of the tube..

Ice scoop

An ice scoop is necessary to clean the hole after it is drilled with an auger and as it is chiseled with a spud bar. If you do not bring a scoop, nothing else you have with you but your bare hands will fit down in the hole. Putting your bare hands in ice cold water to clean the hole is a foolish thing to do. The water will suck the life out of them. Your hands will swell and become numb and useless. Using bare hands to grab a fish is popular on television and in videos, but this is not advised, either. Just pull the fish you catch through the hole with the rod and line. You raised the fish from the bottom of the lake, so another foot out of the hole is not going to knock it off.

An ice scoop is inexpensive. A spatula or spoon with holes in it is probably in your kitchen and equally effective.

Auger

Hand augers are nothing more than a one-piece brace and bit, about 4 feet in length. Get the narrowest size, normally 4 to 5 inches in diameter. You want the narrowest size because wider augers are much more difficult to use. You do not need a large hole to pull a bluegill through the ice.

During the off-season, keep the auger blades covered when stored; it is a safe way to store an auger and protects the edges on the blades. Cover the blades when transporting between sites; you will protect yourself and others from getting cut on the exposed blades.

Ice fishing begins after the holes are drilled. Nothing is worse than trying to drill holes with a dull auger. You will crank yourself into exhaustion with little to show for your efforts.

Rod holder

A rod holder for the second rod is a good idea, but not necessary for success with two rods. Propping one rod above the ice keeps the reel clear of snow and slush. You can look over at the rod in the holder while you jig with the other. You may want a holder for each rod.

Bucket

Large buckets are versatile. You need something to carry tackle. Then you need something to carry fish, and you may want something to sit on in the meantime. A plastic, 5-gallon bucket works well. A flat-bottomed, plastic sled with a raised lip to go with the bucket is even better.

Safety

Safety is the responsibility of each angler. Call the local fish and game office or sheriff's department for ice conditions. Then, cut a hole in the ice at the shoreline before walking out on any lake or pond. The thickness and quality of ice along shore should give you a relative idea of what you will find off shore.

Ice fishing rods are short and sensitive. The rod pictured is suspended in a rod holder, which keeps the reel out of the snow and allows the angler to watch the rod tip from a distance while jigging another rod.

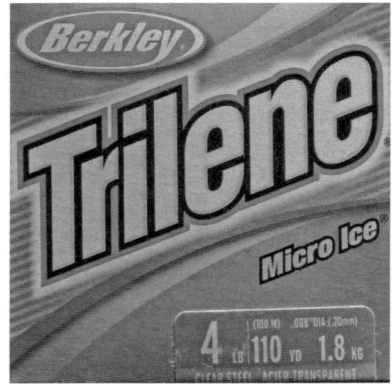

Monofilament line designed for ice fishing is manufactured for high-tech presentations, cold temperatures in particular. Because jigging is gaining in popularity, lines such as Trilene Micro Ice are being designed for less stretch, an advantage when jigging.

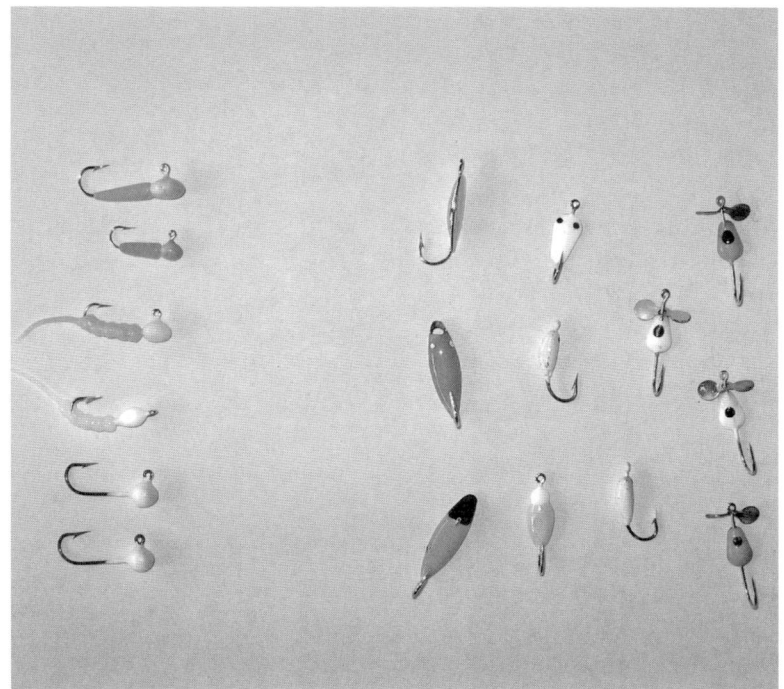

Most jigs used hang horizontal from a line, as pictured on the left. The hook eye comes out of the top of the jig head and hangs parallel to the bottom of a lake. Ice jigs are often vertical, as pictured on the right. The hook eye is not bent in a 90-degree angle. Not all ice jigs are vertical, but many are to facilitate vertical jigging. Whereas horizontal jigs are popular with warm weather fishing and are retrieved through the water with pulling and reeling motions.

Horizontal jigs offer an advantage in that the hook hangs on the same plane as the movement of a bluegill. As it swims to the bait, the jig is easily inhaled straight into the mouth. A vertical jig has to be turned upward by the fish to fit in the mouth. This is easily accomplished with a hearty inhale, but a hearty inhale may be hard to come by when the water is ice cold. Cutting out the extra step should lead to more hooksets connecting with the fish.

Ice fishing floats do not have to be specific. You can use the round floats you use in warm weather or smaller versions of these same floats. Keep in mind that the smaller the float, the less weight needed, to a point of diminishing returns when you no longer have enough weight to keep tension on line stiff from cold temperatures.

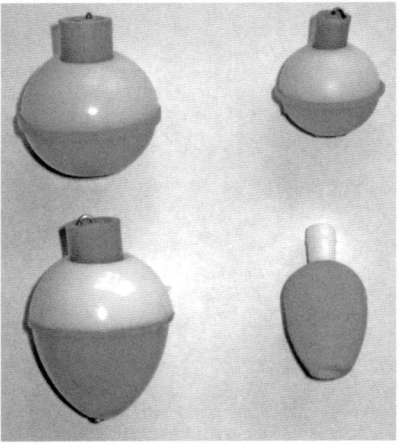

Slip floats freeze at the narrow hole in the center stem. You can modify a plastic float to become a slip float (below left inset) by propping open the brass hook and running the line through it. Another option is a float designed for ice fishing such as the Ice Buster Bobber. An innovative design, the long slot reduces the chance of line freeze. Easy to install, a notch in the head clips to the line (lower right inset). Cut the foam to match the weight of the sinker or split shot, another great idea.

Here is a modified plastic float in action. Plastic floats, those plain old spring bobbers, are versatile and inexpensive. You do not have invest large sums of money to catch fish during every season of the year and do it well.

Start fishing with your bait about 1 foot above the bottom. The easiest way to set your float is to attach a depth finder. Clip the depth finder to the shank of your hook with your tackle fully rigged and ready to fish except for the bait. Let the weight pull the float underwater, then move the float up or down until it is about 1 foot under the ice. Reel the line in, remove the depth finder, add bait, and start fishing. If you do not get strikes at the bottom, set the depth to about two-thirds the distance to the bottom and try it there. If you do not get any strikes at two-thirds the depth, then try a depth about half the distance to the bottom. If you do not get any strikes half the way down, move to a new location and return to this one later, if necessary.

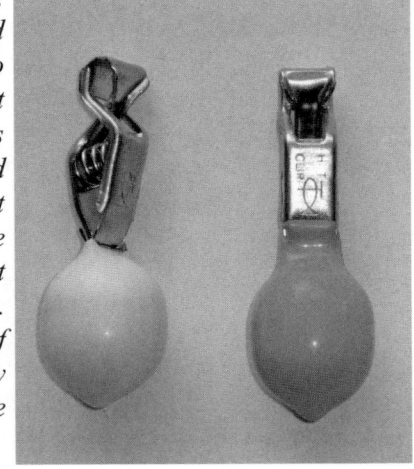

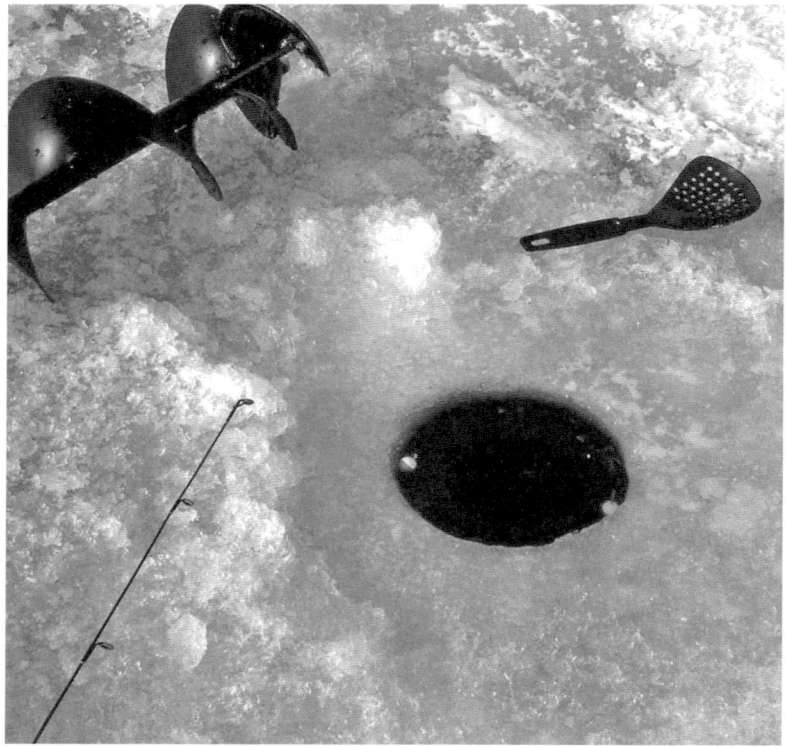

When choosing a hand auger, buy the smallest diameter you can. The size above is designed for the Great Lakes and is much larger than you need for bluegill fishing.

Drilling holes will be much easier with a smaller diameter. And, match the length of the shaft to your height. You do not want an auger as tall as your shoulder. If the free-floating handle (large plastic cap at the upper end) is too high, you will not be able to bear down on the auger. Drilling will be inefficient and the work exhausting.

Augers do not come in assorted lengths, just diameters, but may vary in length by manufacturer. If you want more mechanical advantage, remove the shaft from the cutting end. Cut a few inches from the shaft with a hacksaw.

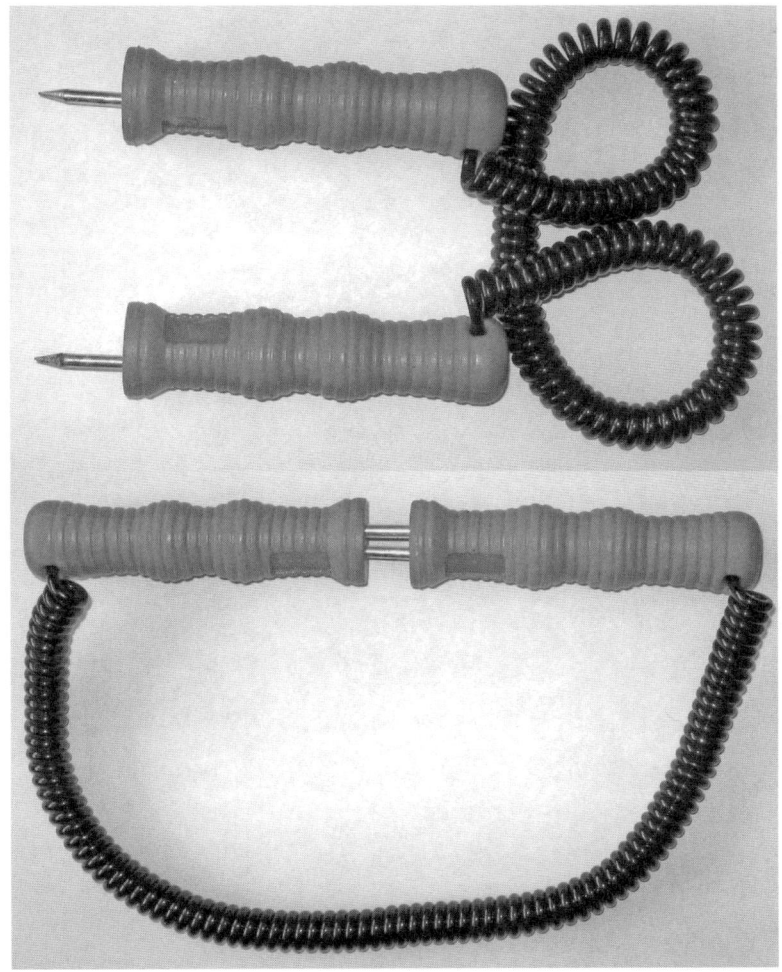

Ice spikes are for pulling yourself up and back on the ice if you break through. Designed to be carried around the neck, many anglers make their own with nails and pieces of a shovel handle. The best piece of safety gear an angler can depend on is knowledge. Know the ice conditions before you go, cut a hole at shore before stepping out, and never go alone.

Tip-ups are unattended devices consisting of a spool of line supported underwater to prevent freezing. When a fish takes the bait, a flag tips up to signal the strike. Tip-ups are fun to watch and easy to use—no need for floats or jigging action. Check fishing regulations in your area to determine the legal number of lines allowed in the water while fishing.

Crappies are often caught when fishing for bluegill, no matter how you are fishing. What does matter is the bait you are using. Crappies like minnows and artificial lures more than night crawlers and garden worms. This crappie was caught on a wax worm.

Breaded fillets are easy to prepare and delicious.

11

COOKING

Bluegills are one of the finest tasting fish in freshwater. The meat is firm and sweet. How the fish is handled after it is caught will determine how it will taste when cooked. How it is cleaned and cooked is equal in importance.

Whether you are a novice or veteran, learning how to care for and then cook the fish can only be explained to a certain extent. Only practice at cleaning and cooking will enable you to get the most from your catch. Three steps to delicious fish will be discussed as follows:

Handling
Cleaning
Cooking

Handling is just as important as cleaning and cooking. Make as much effort with each step to insure the taste will be best possible.

HANDLING FISH

Proper handling is paramount to fresh fish. The best way to handle fish after the catch is to cool the meat immediately. The sooner the better, the colder the better. Many meals are ruined through poor handling. Once the fish dies it must be kept cold until it is cleaned and cooked or frozen. Handling the fish in the field is where many people fail. Proper handling is not difficult; it just takes planning and some additional effort.

The best tasting fish are caught with that purpose in mind. Just as you would prepare tackle, select bait, and choose a location to fish, you should prepare for proper handling of the catch. Warm temperatures are the worst enemy of fresh fish. The meat is too delicate; a matter of minutes in the warm air or warm water can ruin your next meal from the beginning.

Coolers

Time was when coolers were metal containers called ice chests. Putting bluegills on ice as soon as caught is the best way to preserve the meat and flavor. Nothing beats ice-cold fish bound for the table.

A live well may be another option for some anglers, but you still have to ice the fish down for the trip home. With a live well you can release fish if you do not want them. Putting fish in a cooler filled with ice is a one-way trip.

Ice may not be practical for everyone or every fishing trip. You may not want to deal with the hassle of carrying ice if you will be fishing a long distance from your vehicle. You may not want to deal with ice for other reasons. Ice is not mandatory to handle your catch properly.

Ice cubes last longer than ice chips, and both can be added to the cooler and used as is. If the ice is a couple of large blocks, break the ice into smaller chunks to more effectively chill the entire cooler or add enough water to cover the bottom by a couple of inches. The water will super cool the fish, otherwise fish not in direct contact with the ice blocks will take longer to cool, in particular when the cooler is filled with fish.

Baskets and stringers

Wire baskets are the next best option in handling fish after ice and live wells. Baskets are live wells, only not as convenient. A basket hanging from a boat snags in brush and other obstructions, and has to be pulled from the water when you move. If you move a considerable distance on a warm day, the fish will be dead before you arrive. Moving several times even for short distances may have the same effect.

The worst thing that can happen to fresh fish is to have them stay at air temperature for any length of time. The flesh becomes mushy,

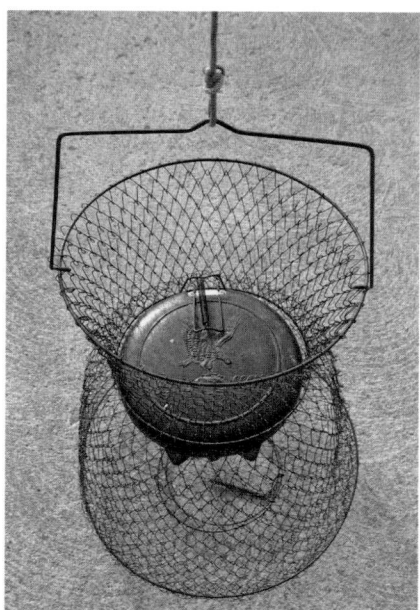

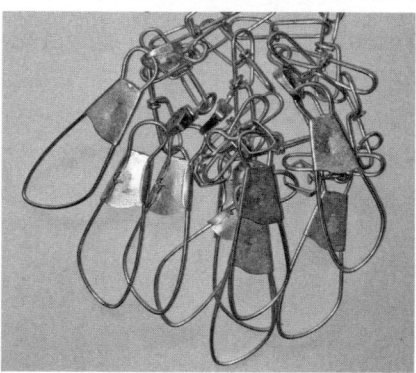

A chest (cooler) filled with ice is the best way to keep fish fresh after you catch them, but also the most inconvenient when not fishing from a boat. A wire basket (left) is the next best way to keep fish fresh by keeping them alive until you leave. Plus you can release fish you no longer want to keep. A basket is ideal for fishing from a boat and shore. The chain stringer (below left) is a lightweight option that will kill a bluegill eventually. The rope stringer (below right) is the least expensive and the most lethal. A small-mouthed fish such as a bluegill will not last long on a rope stringer, spoilage will follow immediately thereafter.

never to regain firmness. And the flavor? Think mush and you'll be close.

If choosing a fish basket, look for one with a hollow lid made of plastic. A plastic lid will float the basket. A floating basket provides two benefits. First, it does not collapse on the fish inside and kill them. Second, you do not have to pull the basket out of the water each time to drop a fish in. A door on the bottom of the basket is another option. The basket can be turned upside down to remove fish from the other end, which is helpful when bluegills become stuck in the mesh by their fins.

Chain and rope stringers are used to secure the fish you catch. Small fish, such as a bluegill, die on a stringer in a short period of time. A stringer is not a temporary holding device, so be sure of the fish before you string them. A wire basket is more effective than a stringer and just as portable.

Harvesting your catch

I encourage all anglers to keep the bluegills they catch, but not every fish. Choose wisely and keep only what you can use. Bluegill populations are self-sustaining fisheries throughout the United States. Keeping the smaller fish (less than 8 inches) will maintain a healthy population. Releasing the larger adults is the surest way to protect the population from over harvest. A 7-inch fish is not more difficult to clean and tastes sweeter than the older and larger fish.

The smaller the body of water, the more careful you should be harvesting fish. Ponds are susceptible to over harvest more than lakes. My training in fisheries management in the late 1970s suggested ponds under proper management could withstand harvesting 40 to 60 percent of the biomass of bluegills each year. If there is any doubt about whether to keep a fish, let it go instead of holding it in a live well or basket. You can catch more bluegills.

Fillet knives come in 4-, 6-, 7½-, and 9-inch lengths, with fixed or folding blades, some with removable blades. Electric knives are becoming popular, often used to clean large quantities of fish on a regular basis. A 4-inch blade may seem appropriate for bluegills, but a 6-inch blade has greater versatility slicing the tall profile.

Putting bluegills on ice as soon as caught is the best way to preserve the meat and flavor. Nothing beats ice-cold fish bound for the table. Notice the dark color, a sure sign of fresh fish that have been handled properly.

CLEANING FISH

Fish taste best when filleted and skinned before being cooked. Filleting removes most if not all of the bones. You should not have to work to eat the fish; do your work ahead of time through cleaning. Boneless fillets are the best way to go.

Filleting the fish removes bones. Fish cooks faster when the bones and skin are removed. Plus, bones in the fish can ruin the meal for people who do not normally eat fish or are squeamish.

Some fat and oil found under the skin are removed when the fish is skinned. Fat and oil make fish taste stronger—you know, the dreaded fishy taste.

Skinning and filleting the fish may seem difficult at first, but with practice you will choose to fillet and skin all the fish you catch, not just the bluegill. As soon as the fillet is removed and skinned, place it in a bowl of ice cubes and water. You want ice water rather than just ice to keep the entire fillet moist as well as cold.

Keep the fish on ice until cleaned, and then put it back on ice after cleaning. This is the best way to protect the flavor and firmness. Fish in ice water can stay that way overnight for cooking the next day. All other fish should be frozen as soon as the cleaning is finished.

Freeze the fish in water. Submerging fish in water decreases the chance for freezer burn. Plastic containers and freezer bags can be filled with water and frozen with the fish, but use only enough water to cover the fish. Make sure the fish placed in a bag are still covered after you set the bag in the freezer. The bag will change shape after you let go, sometimes leaving fillets above water if there is open space in the bag.

Regardless of how the fish are cleaned, keeping the fish ice cold through the entire process is the last and most important step. All your hard work can spoil if you let the fish warm to room temperature at any time between harvest and cooking.

COOKING FISH

Frying is a popular and tasty way to prepare bluegills. The secret to deep-frying is timing, putting the fish in when the oil is hot and then not for too long. The fillets have to be removed the minute they are cooked.

Breading

Breading creates the foundation for taste. You can make yours to your liking, from flour, breadcrumbs, cornmeal, saltine crackers, corn flakes and other cereals. Crackers and cereals need to be ground into meal. Grind the batter with a rolling pin or wine bottle and a gallon size Ziploc bag. The bag will contain the meal and then can be used to coat the fillets after being dipped in egg.

You do not have to roll your own breading; many commercial

You can roll your own breading. Start with your favorite crackers as the main ingredient and season if you wish. Saltine crackers are good right out of the box. Adding cornmeal will change the texture and taste.

preparations are available. Another secret to good-tasting fish is starting with breading that suits your taste, before you cook the fish. Wet your finger in your mouth, stick it the breading, then taste the breading. If the breading does not appeal to you, why make your fish taste the same way?

So when making your own breading, taste often until you create the flavor that suits you. If you are adding salt, pepper, paprika, herbs, and any other seasoning, make your breading taste good before adding the fish. Once the fish is added, it is too late to control the taste, in particular after the fish is cooked and the coating is sealed.

Batter

Batter wets the fish to accept the breading. Egg and milk are the most common ingredients for the batter. The batter is the glue for the crust that forms during the frying process. An important step not to forget is precoating the fillets before the batter dip. This step dries the fillets. Think of when you glue something; if the pieces are wet, the glue cannot do its job. The batter is glue for the breading; when batter sticks to the fillets, so does the breading.

The precoat is nothing more than flour. The reason for this step is simple; the fillets are wet from ice water, and dropping the wet fillet in the batter is like trying to glue something that is wet. Batter will not coat wet fillets to the same degree as dry fillets. Pour about one cup of flour into a Ziploc bag or Tupperware container; add fish and shake to thoroughly coat. Now you are ready for the batter.

- Typical batter
    1/3 cup of milk
    2 egg yolks

Mix the egg and milk with a fork in a bowl or Tupperware container. Drop the floured fillets in; turn several times in the batter. Remove the battered fillets and drop in the breading. You can use the same bag or container that you used to precoat the fillets in flour. Pour out the excess flour before adding the breading.

A closed container is recommended for the breading because you want to shake the fillets with some force to embed the breading.

Batter is the glue that binds the breading to the fillet. The batter sticks to the fillets if the fillets are dry, which is as easy as dropping them in flour before the batter.

Tupperware containers are ideal for frying fish. With the lid on you can shake the contents to mix the ingredients. Then you can use the container to coat the fillets with the breading.

Deep-frying

Heating the oil is important. Fish should not be added to the oil until it has reached the cooking temperature. When the oil is not hot enough to cook, the fillets soak up oil, making them greasy. The taste is not desirable and the excess oil is not healthy. Hot oil seals the batter, making it crispy and golden brown.

The temperature range most recognized for deep-frying is 350-375 degrees Fahrenheit. Do not worry about the exact temperature; just make sure the oil is fully heated before adding fish; it should be hot enough to bubble when a bit of batter is added.

Do not let water from your hands or utensils fall in the hot oil. Water will pop and possibly send searing hot oil in your face. If you do not wear eyeglasses, safety glasses are recommended if you have not deep fried fish before or are not familiar with the fryer you are using.

Depending on the heat source and the type of fryer, the oil may take several minutes to reach cooking temperature. If you do not have

A deep fryer with a fixed temperature setting is safer than an open pan on a burner. A pan of oil should never be left alone on a stove. Oil will ignite when it becomes hot enough. A fryer with a thermostat is the most versatile.

a thermometer or thermostat on the fryer, drop breading in after 5 minutes, then repeat this step every 2 minutes thereafter until the oil boils vigorously and the breading floats to the surface.

Carefully add the fillets to the oil and cook for about 3 to 6 minutes. Be prepared for the oil to rise up in bubbles when you add the first few pieces. Because of this, you want a deep pan to fry in; do not fill the pan more than half full. Hot oil can become extremely dangerous if not treated with care.

Do not leave a pot of cooking oil unattended or on the highest setting of your stove. Once the oil reaches frying temperature (a cooking thermometer is recommended), turn the heat down to maintain a steady temperature while frying. A deep fryer with a fixed setting is safer than a cooking pan on a burner and no thermometer. A fryer with a thermostat is the most useful.

Cooking time depends on the temperature of the oil and the thickness of the fish. Fillets float when cooked through. Turning the fish with high-temperature spatula will release the fish at the bottom and allow the cooked pieces to float to the surface. Do not overcook; even bluegill will become dry and lose some of the delicious taste.

Oil is something to consider before cooking. Vegetable shortening, butter, and lard are often using for frying. These products have lower smoke points (the temperature that an oil begins to smoke) than other oils such as peanut oil. Peanut oil has a high-smoke point, but is expensive and not mandatory to cook a batch of delicious fish. Safflower is another option, as well as corn, canola, and soybean oils. Most grocery stores carry a substantial selection of cooking oil.

Refined oils, those collected with high heat, have higher smoke points than unrefined oils. Refined oils have less color and taste than unrefined oils. Unrefined oils are collected by cold extraction; they are darker in color and stronger in taste.

Melted butter and lard are not the best oils to use when deep-frying. You have options available at the grocery store.

Fried bluegill fillets are delicious. If you have not tried them, you have no idea of what you are missing.

Baking

Bluegill can be baked with the same ingredients used for deep-frying. The breaded fillets are cooked in the oven instead of fried in oil. The taste may seem drier when compared to frying.

- Spray nonstick coating in a baking pan. Butter or olive oil can be used instead of the nonstick spray
- Preheat oven to 425 degrees
- Bake about 15 minutes or until meat flakes off with a fork
- Place fillets on paper towel for a few minutes before serving

A jig is an effective search lure not often used for bluegill. It is versatile enough to be fished high and low and anywhere in between. The upturned hook is less likely to snag; you can add scented plastic for more attraction; if you like to tie flies, you can dress your jig with hair and feathers.

Notice the natural camouflage of the bluegill against a dark background. You can tell the condition of the water and sometimes the color of the bottom by looking at the color on a fish. A fish living in clear water with a dark bottom will look as the one pictured above, with dark colors, especially the olive black on the back.

12

QUICK TIPS

Some of the following information has been presented in earlier chapters, but is important enough to be repeated.

ROD AND REEL

Use what you have. An expensive investment is not needed to catch bluegill.

Rod: Medium-light to medium powered models will work for most presentations. Length is also a matter of preference for general fishing conditions, 5'6" to 6'6" rods will work well for bluegill. If you want specific results, you need specific lengths and actions in your rods.

    A longer rod allows you to use lighter line. A longer rod also offers greater line control and casting distance. Line control and distance are not always desirable for some presentations. A shorter rod will well for vertical presentations such as live bait from a boat or dock.

    A faster action gives greater performance with artificial lures; it bends mostly at the tip. A slower action is more suited to using live bait because more of the rod bends, cushioning the line during a cast and hookset.

    An ultra-light rod with ultra-light line makes for great fun, but is limited to short distances and shallow depths. The whippy rod cannot set a hook past a certain distance because both the rod and line flex too much.

Let experience guide you to when considering new tackle. You should use what you have until you identify specific needs. Time spent fishing is the best way to learn what your tackle needs are without investing money on what you don't need. If you are just starting out, go with a medium-actioned 5'6" to 6'6" rod.

Reel: The reel should match the rod. A rod has the recommended line weights marked on the blank above the handle. A reel is designed to perform within a range as well. This means you will want a reel using medium-light to medium weight lines for most bluegill fishing.

Spincasting: Sales records indicate people buy more spincasting outfits than any other style. A spincasting reel has a closed-face (nose cone). This is the simplest outfit to use. Tangles are few and performance is ideal for many anglers, especially those not familiar with tackle. Spincasting tackle should be your first choice if you are just starting out.

Spinning: High performance comes with a price. Spinning reels can easily tangle, but the extra effort is worth it. Spinning tackle needs adequate tension on the line when reeling in or loops will form and tangles will soon follow. Light tackle, which is most often used for bluegill, may not maintain sufficient tension to prevent loops. When loops form, line must be stripped from the reel, sometimes the amount that has been cast. Hold tension on the line with one hand as you reel in the stripped line with the other.

LINE

Use premium monofilament line. High quality line is necessary, especially when fishing in plant beds and around brush. Braided lines are not necessary, but can be used if you use a piece of monofilament (leader) to attach the hook. The leader can be made from specialty line such as tippet material used for fly-fishing, which is high-quality line, or from the monofilament line on another reel. If you take line from another reel, make sure it is a fresh piece by cutting off several feet before cutting the leader section. A leader does not have to be long in length. *See also* Chapter 3 for details on making leaders.

4-pound test: Light line is more effective because the diameter is not as easy to detect and the presentation is more natural. Light line requires more care. A couple of feet of line may have to be cut off after pulling hard on a snag. Check the end of your line after getting snagged or pulling through cover. Run your thumb and forefinger down the last couple of feet to the hook. If the line feels rough, cut off the section and retie as often as necessary. Cutting the line off at the end and tying a fresh knot maintains strength.

6-pound test: If you want to stick with one line, this is it. You can stop a canoe floating gently downstream when snagged with 6-pound test. Stopping a canoe with your line has nothing to do with fishing for bluegill, the example describes how strong premium monofilament is, which is much stronger than most people think. With 6-pound test you can fish in brush and rocks with excellent results. In a plant bed the weeds give before the line. Cattails may break the line, but not every time. You should keep the line fresh by cutting off the end as you do with 4-pound test, but not as often.

8-pound test: The heaviest line you need is 8-pound test. A spincasting reel is often filled with 10-pound test, which is will work in most fishing conditions with favorable results, but replace it with 8-pound line the next time you service the reel. New anglers will benefit from a stronger line such as 8-pound test. You do not have to stop them from fishing to check the line and retie knots. And you do not have to tell them to be careful when snagged or pulling in a fish tangled in the weeds.

HOOKS

Buy the best hooks you can afford. You do not want to save money on hooks. The hook is your link to the fish. If the design is poor, you will miss strikes. If the quality is poor, the point will roll or break off when you least expect it. After missing several fish, you may figure out why, then again, you may not.

Most people use hooks in sizes too large for bluegill. A bluegill has a tiny mouth; large hooks make it harder for a bluegill to take the bait. Long-shanked hooks are also popular, as if to prepare for easy removal of swallowed hooks. Do not plan for hooks being swallowed; it kills the fish and anglers rarely keep every fish they catch.

Circle hooks are recommended for live bait. Circle hooks are virtually swallow proof, which makes this design ideal for fishing with live bait. Some applications with scented plastic (grubs and tubes), such as those still fished under a float like live bait, also warrant their use.

New anglers who have not yet learned to set the hook will benefit from using circle hooks. The fish hooks itself as it swims away. Another benefit to using circle hooks is not dealing with a swallowed hook, a distinct advantage when teaching new anglers. The bloody mess of a swallowed hook can be enough to ruin an outing or the whole interest in fishing.

Short-shanked designs are excellent choices for small bait such as wax worms and garden worms.

Anglers who use crickets often desire the Aberdeen style, a more traditional hook made of thin wire. Both the styles will work. When high-quality hooks are used, which hook style is best depends on the bait used and personal preference of the angler. As with any other tackle, use what you like, you will fish with confidence. It is how you use the hook that makes a difference. Buying high-quality hooks will take most of the guesswork out of choosing.

JIGS

Jigs are another option for anglers using live bait and plastic bait. A jig can be still fished under a float or retrieved as a lure. You can use a jig in any manner that a bare hook would be used. A jig is a bare hook with the hook eye bent at a 90-degree angle and weight is added at the 90-degree bend.

Serious anglers, especially those who fish through the ice, know well the effectiveness of using jigs for bluegill. But the rest of the fishing population may not take full advantage of using jigs with and without live bait.

LURES

Lures (plugs) are fun and effective if small enough. Several lure makers offer miniature lures for panfish anglers. Plugs at times are too effective; the treble hooks make removing a bluegill difficult. Hemostats or needle-nosed pliers are recommended to remove the hooks from such a small mouth.

Floating minnows are ideal for fishing over plant beds. Shallow runners can be used as search baits to find the bluegill in the back of a cove or bay. Sinking minnows are good for fishing deep along the edges of a plant bed. They can also be used as search baits at the mouth of a bay or cove as it meets deep water.

Fat-bodied plugs are also popular and effective. The broad design creates pressure waves different from those made by thin-bodied plugs. One design is not better than the other. Do not worry about choosing the right plug. What works well in one lake may not do as well in another lake because the available forage and habitat will not be identical. What is worth you concern is the retrieve you use. Do not just reel in the plug. Change retrieves to include stopping and jerking. Let the plug sit after each cast. Then stop every so often during the retrieve.

SPINNER BAITS

Safety-pin spinners can be your most productive bait for several reasons. Spinner baits can be fished at any depth; the twirling blade adds sight and sound to the presentation; the angled arm posed over the jig makes it less prone to snagging; live bait can be used instead of scented plastic and feathered dressings.

Traditional in-line spinners are also effective, but the weighted treble hook snags everything it touches. This design can prove a disaster around wood and vegetation. Use in-line spinners over rock bottoms and in open water as search lures when you do not know where the fish are and do not have a plug that will do the job.

ICE FISHING

Look for bluegill in the areas where you find them during the fall. Most of the time this will be in water somewhat deeper than in the spring-spawning period, but still associated with vegetation, wood, rock piles, and other such structures on the bottom. A bed of sand on the bottom adjacent to these areas of structure adds to the attraction. Insect larvae will be living in the sand, which provides a food source next to the cover.

The bluegill will move around during the day as temperatures rise. Often the successful strategy is to stay relatively close to shore but move about until you find fish—if you do not have electronics. Staying close to shore keeps you in the littoral zone, the productive area of a lake, where sunlight penetrates to the bottom. This is a good area to be at, where food and cover are in close proximity. Water temperature plays a key role in ice fishing. The littoral zone is the place to be. The sunlight will be hitting the littoral zone, not with the same intensity or at the same angle as during the summer, but the sun will have a marked effect if snow cover does not block the rays.

Use wax worms for bait; they are easy to find and inexpensive.

Maggots work when wax worms don't, but may not be as easy to find at smaller bait shops. Larger lakes will support full-service bait shops.

BAIT

Wax worms are the perfect bait for bluegill. Bite size and soft, they are also sweet scented from eating beeswax. Wax worms are effective 12 months of the year in warm water and cold. Use 1 or 2 wax worms at a time; hook them once near the head.

Garden worms are more effective than night crawlers if for no other reason than size. The smaller garden variety of worm can be inhaled with much less effort. Both worms are members of the same scientific family and look identical to the naked eye. Night crawlers will work, but the worm is so large that even an adult bluegill cannot fit it all in its mouth. The result is missed strikes and bitten-off worms. A bitten-off worm loses its appeal in pressured waters, as does a night crawler broken into pieces. If you are fishing in water that does not receive much pressure, night crawlers and pieces of night crawlers will start a feeding frenzy.

Regardless of which worm you use, hook it once. Do not thread the worm on the hook or hook one through the collar. Hooking a worm through the collar, even once, will kill it. A dead worm on the hook has no appeal when compared to a lively one wriggling around in front of the fish.

Crickets are probably the favorite bait of the South. Once popular in many areas, the availability and ease of keeping wax worms have made crickets less common in some areas. Crickets are every bit as effective as wax worms. A couple of drawbacks are availability and lack of staying power. One bite and the cricket is gone, fish on or not. A wax worm stays on the hook a little better than a cricket. Hook a cricket once as you do other baits.

Leeches make outstanding bait for big bluegill simply because the little bluegill will not normally mess with a leech. A leech has thick muscle and stays on a hook forever it seems when compared to other bait. Availability is limited to the larger bait stores if you cannot catch some yourself.

Other bait includes insect larvae from the local streams or your yard. Grubs pulled from under the bark of dead trees are also good. Lunchmeat and white bread provide effective bait as well.

The last bait to consider is scented plastic. A scented tube or grub can be fished with or without action, on bare hooks or jig heads. Unscented plastic can be used but an active presentation will be necessary to draw strikes.

RIGGING FOR BAIT

A float, sinker, and hook are required rigging for bait. Simple yes, but oh so effective. When nothing else will work, bring out the live bait and watch your luck change.

Set the float to present the bait from one-half to two-thirds the distance to the bottom. If the fish are actively feeding and hitting the surface, move the float up and fish in the upper one-half of the water column.

Rigging is more versatile when a slip float is used instead of a fixed float. A slip float can be adjusted to any depth just by sliding the stop up and down on the line.

A fixed float can be moved as well, but it has a limit to the depth used. A fixed float set deep becomes impossible to cast because too much line hangs from the rod.

Slip floats allow unlimited depths without interfering with the cast. A slip float slides down the line before a cast is made, making any depth possible.

Float size should reflect the length of the cast. If a long cast from shore is needed, you will need a large float. The combined weight of the float and the sinker will make the cast possible. Cast farther than the target, which will reduce the disturbance from splashdown. Then, reel the bait to the target. If a short cast is possible, use the smallest float needed to cast the bait. The disturbance will be less likely to spook the fish.

PLANTS

Plant beds create prime habitat for bluegill. Submerged plants, those growing underwater are the most attractive because they grow deeper in the water than the emergent plants such as cattail, bulrush, and floating-leaf species. Submerged plants are more difficult to recognize, especially from shore. If you cannot find a submerged bed, then try the emergent plants you do see. Start fish near the plants in the deepest water.

    A plant bed provides the food and shelter a bluegill needs to survive, both in the same place. A plant bed is crawling with aquatic insects and invertebrates. A bluegill can find all it needs to eat in a healthy bed.

    The stalks and leaves offer shelter from predators such as the largemouth bass and northern pike. The thin profile of a bluegill is designed to maneuver between the stalks, for escape, and for feeding. A bluegill will find more than enough to eat among the stalks; there is no need to leave the safety of the plant bed. Plants provide shade and a comfortable place to rest. A dense bed acts as a screen, filtering out silt and suspended sediment.

    Before making a cast, look for pockets and holes. Changes in the bed will attract the bluegill. Corners and pockets along the edges of a bed create ideal pieces of cover in the habitat. Pockets of open water above the bed between the tops and the surface are equally attractive. A hole in an otherwise solid portion of the bed is sure to hold a fish or two.

RIGGING TO FISH OVER PLANTS

Fishing over a plant bed requires a shallow presentation. The rigging begins with a casting bubble or fixed float, used as weight to propel the cast. No other weight is used. The round-plastic float is reversed to serve as a casting bubble. Reversing the float make it more streamline to be pulled through vegetation, if necessary.

Attach a swivel to the running line, in front of the float or casting bubble. Reeling in this rig causes the line to twist. A swivel reduces twist in the line.

Then attach a leader to the rear of the float or casting bubble. For the leader use a 3-foot piece of line from your reel. You can also use fly-fishing tippet, including fluorocarbon. Fly-fishing tippet is thin and strong. Fluorocarbon is less visible under the water. Fish living among the plant tops do not receive the same amount of pressure, so fluorocarbon line is not required, but it may increase strikes because it is less visible. Fluorocarbon is expensive and does not have the same properties as monofilament. The line on your reel is recommended first, then the fly-fishing tippet if you seek higher performance, and fluorocarbon if you want higher performance still.

A bare hook or fly is attached to the end of the leader. The hook with bait or fly rests near the surface. A twitch every couple of minutes is enough to make the presentation irresistible to those fish isolated among the plant tops. They are waiting for just such an opportunity.

APPROACH

Catching an adult bluegill presents many challenges. Your approach should focus on finding one at a time. You will catch more and larger fish by making each cast with purpose. Pick a target before making a cast. Bluegill are not the fish most people think they are, especially the adults.

THE IMAGE OF YOUR FLY

Image is what makes a bluegill strike a fly, an optical illusion; no scent and little if any sound are involved. The goal when fly-fishing is to make a fish react the moment it sees a fly, to do what it takes to trigger a strike.

A reflex governed by instinct does not fail easily in water receiving heavy pressure from anglers. A fish eats and spits out enough foreign objects to recognize what is not worth the effort or what has in the past held a hook.

A fly attracts a strike through visual recognition—when what looks like food acts like it. Aquatic insects move in certain ways, as do crayfish and minnows. Realistic behavior from a fly is subtle, requiring far less movement than most anglers impart by stripping. Unrealistic behavior can kill a presentation before it begins.

Twitching a fly is very effective for dry and wet flies. A presentation should start with no motion. Let a dry fly sit on the water, for a couple of minutes if you can stand it. Then, twitch it, let it sit, and then twitch it again. Continue twitching across the water until you find the fish. When you find the fish, cast back to the spot and let the fly sit, again. Start twitching again, if necessary.

Wet flies can be allowed to sink to the bottom with no motion, *if* you know where the fish are. If you do not know where the fish are or at what depth, you want to start high in the water and work your way down. Otherwise you can spook the fish with the first presentation by pulling the line down on them.

A bite often comes on the drop and detection is tough. Be diligent watching the leader; set the hook if the leader stops or twitches itself. If the fly passes to the bottom undetected, twitch it in. Twitch it in, until you find the fish. Twitch at different depths if necessary, start high in the water and then work the fly slightly deeper with each succeeding depth.

NO TWO ARE ALIKE

A body of water is similar to other bodies of water in that each serves a function in the water cycle, storing water. However, beyond the basic function is where the similarity ends. From the perspective of fishing, one body of water will fish different from the next.

Several factors affect the water. The depth of water, bank angles, amount of vegetation, and bottom cover are some of the physical features that can make a body of water unique. The soil surrounding the water is the source of nutrients feeding the water every time it rains a substantial amount. The amount of trees growing around the water, the land use (urban or agricultural), and the streams and springs recharging the lake or pond are also factors.

So many variables are involved that all you really need to know about a body of water is no two are alike and what works in one will not work the same in another. The fish will behave differently and react differently to bait and lures. You can catch bluegill with the same methods and the same or similar tackle, but perhaps not at the same depths and times.

This tip came to mind when my friend Barry, pictured holding a whale at the beginning of Chapter 6, asked me why the water we often fish (in the picture behind him) does not have a dry-fly bite as his favorite water does.

Three factors often determine the character of a body of water: depth, temperature, and chemical content.

Temperature affects the metabolism. Temperature can speed up and slow down feeding activity. Temperature too warm in the summer can cause feeding to slow or cease just as temperature too cold can do in the winter. Then, when you compare two bodies of water, the temperatures will not be the same in each body. In fact, the temperature will vary within the same body of water, so the fish will be at different stages within the same body of water if it is large enough.

Depth comes into play somewhat connected to temperature because the deeper the water, the more thermal capacity it holds. More water means the temperature changes more slowly, regardless if it is warming or cooling. Deeper water means more habitat and this will also make a deep body of water fish differently from a shallow body.

Chemicals make a difference not so much from a pollution concern, but more because of the presence of necessary nutrients such as phosphorus. One body of water will not contain the same chemical content because the soils will be different. How the land around the water is used (farming or housing for example is also a factor). Soils will have different chemical (elemental) properties.

A body of water with a balanced concentration of chemicals will be productive, creating plankton blooms, which establish the food base and support the bluegill. Some bodies of water will be depleted because of age; others will be overdosed from pollution.

Another feature to consider when comparing is the water clarity. Clear water is less fertile and fluctuates more in temperature because sunlight penetrates deeper and less of the light is reflected. Many features, beyond those mentioned, affect the fish and make fishing in the water somewhat different from other bodies.

To answer Barry's question, I had to ask him a few about his favorite water, which I knew nothing about. When I learned lily pads grew in the pond, I knew immediately the pond was much shallower than the one we fished together. The one we fished was 17-19 feet in the deepest section. Lily pads grow to depths of about 6 feet, on average, often shallower than this.

A shallower body of water has a broad littoral zone, with sunlight reaching the bottom, plants galore. Floating-leaved plants create places to hide in shallow water. Bluegill are willing to grab a dry fly all day long. They may be darting out from the shadows provided by the plants. If the bluegill are at shallow depths to begin with, a dry fly will be a common and reliable pattern throughout the day.

Whereas the other body of water is deep and mostly clear. The bluegill will take dry flies, but not often during the heat of day. Sunrise and sunset, low light periods when sunlight is not so intense, are the best times to present dry flies. As the sun rises, the bluegill retreat to deeper water to escape the intense sunlight. Many other organisms living in the pond will do the same.

WATER CLARITY

Water clarity often follows water level. High water is muddy (turbid) and low water is clear. Plankton blooms and pollution can cloud the water as well, but not to the same extent as high water from heavy rainfall.

Bluegill will be shallow, close to shore when the water is turbid. Turbid water offers a bluegill a chance to feed in secrecy, and opportunities to explore areas not normally visited in broad daylight. If the water is too turbid to see through, the fish can be a rod length from shore. Bluegill move to shallower water to feed on prey washed in with the rain.

Clear water has the opposite effect. A bluegill in clear water is exposed to predators. Deeper water offers shelter, a sense of security. If no cover is available in clear water, the fish will go deeper until they find some.

As a check of water clarity, tie a white cup to a length of line and lower it into the water from a dock or boat. Doing this from shore is not practical in most locations unless done from a steep bank. Lower the cup until you can barely see the white. The distance will give you a rough idea of sunlight penetration.

The distance you measure to the cup is a rough bearing on the water clarity. You do not need an exact measurement to recognize the water as more or less clear. If you recognize the water being more turbid than the time before, you may find the fish shallower this time. If the water is clearer, the fish will more than likely be deeper than they were when the water was more turbid.

For example, if the cup distance is about 3 feet and you are catching fish at 6 feet. You may be able to go to another part of the lake and find an area with the same clarity. You should set your floats at about 6 feet if you do not know where to start.

The cup depth is also a means of gauging how deep or shallow the fish may be on a particular day. Keep good records for reference if are serious about using this method. Indicate in your records the time of day, cloud cover, air and water temperatures.

TEMPERATURE

When temperatures rise, so do the fish. This movement occurs most in the winter, spring, and autumn. Summer temperatures are not as easy to classify because the surface water can become too warm (in the mid 80s) and repel the fish to cooler depths. Sunlight intensity also plays a role in the summer.

When temperatures drop, so does the bluegill. Cold snaps are associated with strong-weather fronts and the bluegill may be instinctively avoiding both. As the season changes, temperatures fluctuate, and bluegill move back and forth with the changing temperature. The distance moved and amount of time spent in one place depends on the weather.

SUNLIGHT

Sunlight has a marked effect on bluegill behavior. The more intense the sunshine, the more effect on the fish. Water clarity offsets sunlight intensity. The suspended particles block a portion of the light. Deeper water offers shelter because the rays can penetrate only so far, not always to the bottom.

On cloudy days you can find fish shallower and closer to shore than you normally do on a bright day. Spawning fish ignore the sun and lack of security for the sake of their eggs.

On clear days in clear water the fishing can be slow. Fishing on bright days in clear water will be better during sunrise and sunset with two active periods in between. (*See also* Chapter 9 for an explanation of active periods during the day.) Putting bait in the face of a bluegill is effective throughout the day, and it requires pinpoint presentations in deep water.

WOOD

A recently fallen tree with small branches still attached is more attractive to bluegill than an old log stripped of its branches. Plunk your bait in among the branches where all the snags are, that is where you will find the fish as well. Inactive fish will not move out of a fallen tree to take your bait, but if you put the bait down among the branches, you will increase your chances of drawing a strike.

Do not spend too much time in the brush. The bite will come fast or not at all. Once you find fish at a brush pile, look for other brush piles at the same depth. This will save time fishing at depths too shallow or too deep for the present conditions.

Wood attracts bluegill throughout the year and especially in a protected cove during the late winter and early spring. A cove protected from the wind warms faster when exposed to sunlight. Plants are not yet growing and the wood provides welcome cover.

WIND

If you are fishing from shore, take advantage of the wind. Wind-induced waves will push plankton and other small organisms to shore. Following them will be the fish that eat these organisms, including bluegill. So when the wind makes fishing conditions difficult, put on larger floats with heavier weights and cast into the wind. Let your bait drift with the wind.

Artificial baits can be used if heavy enough to cast. Using a spin bubble, which is heavy, is a good option for windy days. Below it you can attach a jig, live bait, or soft plastic bait.

If fishing from a boat, try to position the boat perpendicular to the direction the wind is blowing. You can then cast at a quartering angle, avoiding the head-on condition.

ROCKS

When you find bluegill using rocks for cover, the fish are often big. The bluegill can find food and cover between the rocks. Look for piles of rock on banks and points. Rocky banks and points near a dam are ideal places to start fishing.

Rocky banks and points are located adjacent to deep water. Food and cover located next to deep water create desirable habitat. These areas are most attractive when the wind is blowing waves against the rocks in the summer. Zooplankton and other prey of the bluegill will be pushed against the rocks to the waiting bluegill.

KEEP MOVING

Bluegill move about during the day, but the reason you should move is simple; if you are not catching fish, you are not in the right spot. Your bait may be at the wrong depth, but more likely you are not fishing where the fish are. As Tommy Martin said, you cannot catch the fish that aren't there.

Bluegill bite on most days, if you can put the bait where the fish are, which is easier said than done. But, if you keep moving to areas with potential to hold fish, you will eventually find them. Potential areas are places offering food and shelter. Food does not have to be linked to shelter. The two can be separate, but one should be located in the vicinity of the other. What should be close to either the food or shelter is deeper water, not deep water as much as water deeper than what surrounds the food or shelter. The following habitat features are potential areas to find bluegill.

Vegetation: Anytime the plants are still standing, especially when the plants are green and photosynthetic.

Branched trees and brush: Late winter and early spring when vegetation is dead and lying down and when vegetation is absent.

Rocks: Prespawn period, especially attractive in areas where clean sand (spawning ground) butts up against the rocks. Look near dams for these places. Also attractive when vegetation and wood are scarce.

FISHING WITH KIDS

Taking a kid fishing is the most rewarding form of fishing. You may create memories that last perhaps for the rest of their lives. The younger the child starts, the more likely he or she will develop a lifelong passion.

But, the younger you start the child, the more patience you will need to teach and then not interrupt the fun, no matter how much you think they should do things your way. Your way is not their way and the more you enforce it, the more difficult you will make it for them. Trying to hard is worse than just being there for them. Give them guidance—once—and let them go.

If what they want to do keeps them from catching fish, wait until they ask why they are not catching any fish, and they will if they are old enough to talk.

Children enjoy the experience of being out, not only catching fish, as they get older they will want more results. Expectations from the children will vary with age and personality. Some may not care at all about fish after 5 minutes, but will enjoy being near the water and wildlife.

Do your homework before going. Look beforehand for a spot with easy access, open areas to cast, and fish. Show them how to make a cast, and then make a cast with them, and repeat making casts with your help until they can do it on their own. A perfect cast is never needed. If they want to try it without your help, let them. After a few tries, they will either ask for your help or let you help them, again.

Keep them safe, comfortable, and always remember that less instruction is more effective.

DEPTH

Deep water is important for security from predators and the elements such as temperature, sunlight, and dissolved oxygen. But the deepest water around is not so important as water that is deeper than the rest. Bluegill use flat areas, this is where plants grow and trees settle. The flat area with deeper water close by is more attractive than the flat area that never ends. Escape over a greater distance takes more energy and poses greater risk

What this means for the angler: Don't look for the deepest water in the lake and pond. Look instead for flat areas with habitat—plants, trees, bush, and rocks—located closer to water that is somewhat deeper than the rest. These places offer quick escapes when needed and short commutes when wanted.

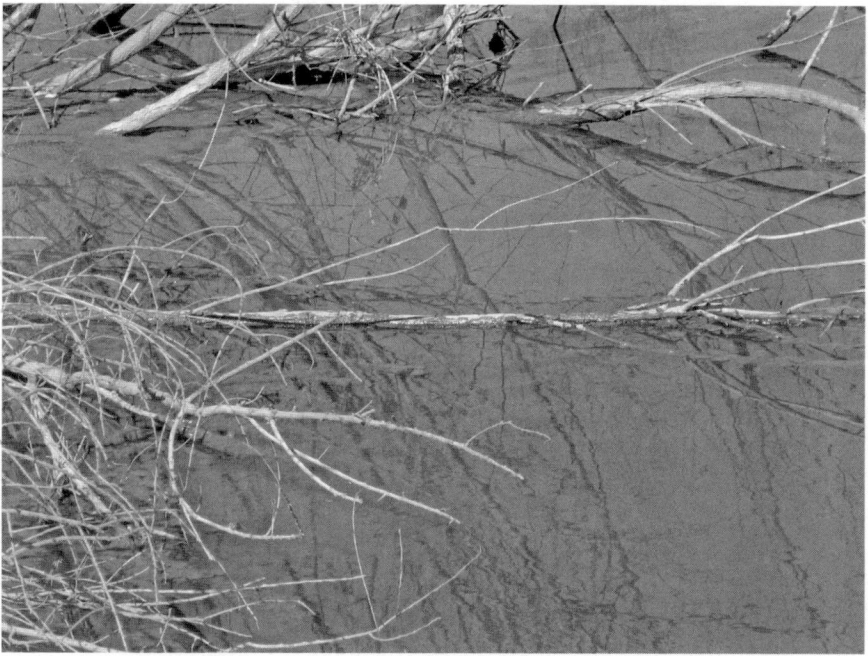

Notice the sand in the upper left corner; the water on the flat is too shallow to attract adult bluegill, brush or no brush. Look for water that is at least 3 feet deep on the flat with greater depth nearby.

HIGH WATER

Bluegill come up from the depths and swim to shore when the water rises after heavy rainfall. Rain pouring down washes worms and other invertebrates from the land into the water. The bluegill know this and go on the hunt for food falling in from the banks.

Rising water also attracts fish to shore in that shoreline habitat is now flooded and available to the bluegill. The bluegill will not only find food close to shore, but some of the shore is now open to hunting. High water is an excellent time to find bluegill close to shore and feeding.

High water after rainfall attracts fish close to shore. The bluegill will swim up from the depths and in to shore to find food. Water clarity determines how close the bluegill will come; the muddier the water the closer to shore the bluegill will be. The ideal location to start fishing is where brush is in the water. Better still is brush where floodwater flows to the lake.

WORMS

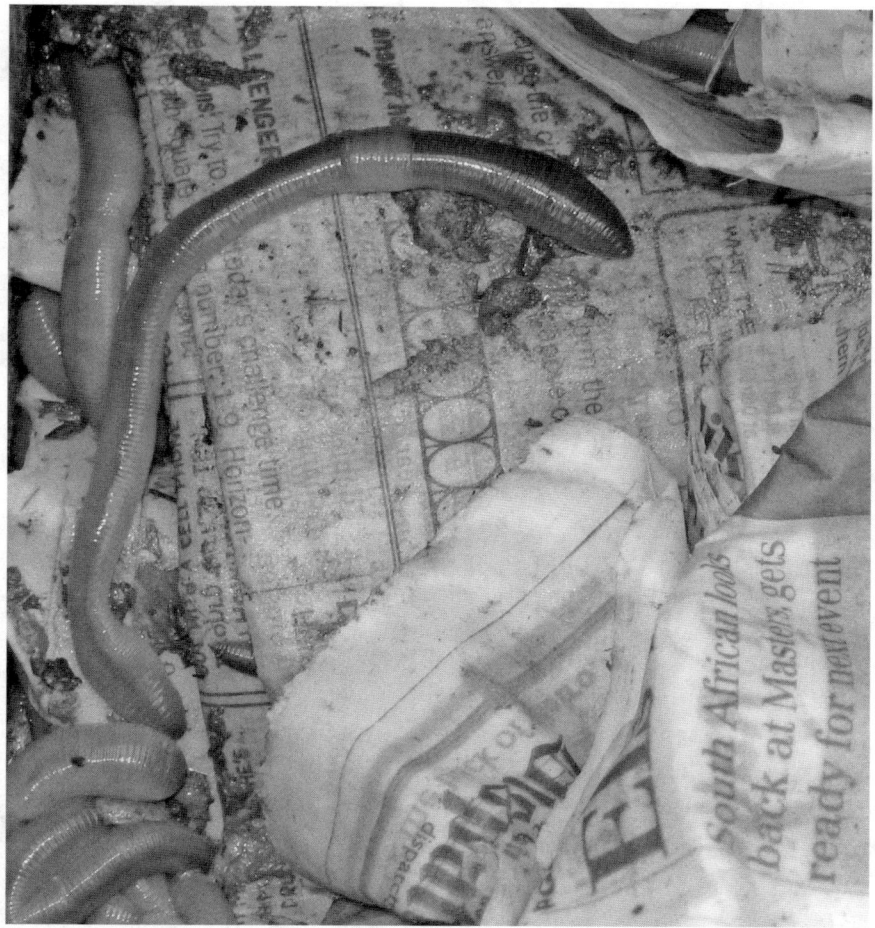

Worms need to stay cool and wet at all times. Store them in a refrigerator for the best results. A basement is the next best thing. Use insulated containers, especially when out fishing. Cover the worms with newspaper, add enough water to keep top layer of paper wet, add water as needed. Sprinkle cornmeal in container for food. Remove the dead and weak ones immediately or the whole batch will soon die. Sphagnum moss can be substituted for newspaper.

FINDING FISH FROM SHORE

The tendency is to run to the best-looking habitat first. Chances are the real fish-holding locations are not always obvious and you may walk right by trying to get to what looks like the good spots.

Pictured to the right and on the opposite page is a clean shoreline with not so obvious features. The brush growing near the water (upper right corner of shoreline) looks inviting, but did not hold fish. The shoreline is lined with smaller rocks and the bluegill were scattered all along the shore where the water was deep enough.

Stay close to shore if you are not familiar with the body of water and do not see desirable features such as vegetation, wood, and rock immediately before you.

Cast parallel to shore where you estimate the water is about 3 feet deep. If you do not get a bite within minutes, cast from same location but somewhat farther from shore to find slightly deeper water, then move down the shoreline.

Make each cast with purpose. Do not cast straight out from shore thinking the bluegill will come. Cast in the order as shown for best results.

TACKLE STORAGE

A tackle bag is more suited for bluegill fishing than a tackle box. The bag is more efficient for fishing on the move, which is often necessary to find the fish. The shoulder strap frees one hand and the large compartments can carry items other than tackle to make your trip more comfortable.

MUST-HAVE EQUIPMENT

Pocketknife: Any knife will do as long as it is sharp and stored with your tackle or carried in your pocket.

Hemostats: No other tool will remove a swallowed hook like a pair of hemostats.

Needle-nosed pliers: You can substitute your hemostats if necessary, but a pair of pliers will serve your needs when working with split shot, sinkers, snap swivels, and most other tackle.

Clippers: Yes a knife will do, but clippers (such as fingernail clippers) and similar tools are just what you need for trimming line. Clippers make the job easier and faster.

Polarized glasses: Protect your eyes from exposure to direct sunlight and sunlight reflected off of the water by wearing polarized glasses. Even on partly cloudy days glasses enable you to see down into the water and identify structure to pinpoint where need to make the next cast. More important than seeing the structure is identifying changes in depth. Often this is the key factor to finding more fish close to shore.

Hand towel: A towel will wipe the dirt and mucous from your hands over baiting hooks with worms and catching fish. A towel is especially useful with new anglers reluctant to bait hooks and grab fish. Show them how to use their bare hands and then wipe their hands on the towel. Do not teach them to use the towel while handling bait or fish; it is just not the way to raise an angler.

Water: A bottle of drinking water goes a long way on any given day. You will fish longer and more comfortably if you are not thirsty and wondering if you should leave because the fish aren't biting.

236 • CATCHING BLUEGILL

ARTIFICIAL OR LIVE BAIT?

The decision of which bait to use is a matter of preference and timing. Artificial bait is fishing in the fast lane, not for every angler, and also not always as effective as live bait. Artificial bait works well in the early morning and late afternoon when fish are most often actively feeding, and at other times of the day when skies are overcast. In water that does not receive much fishing pressure, artificial bait can be good all day long, bright sunshine or not.

If you are fishing in public water, live bait offers the most potential, especially if it is after 10:00 A.M. Live bait is effective under most any condition you may encounter. But, artificial bait may catch more fish because cast after cast is made instead of waiting for a bite. The choice of bait should be based on what fishing methods you like to use, the time of day, and the amount of pressure the water receives for its size.

Bluegill fishing is synonymous with live bait, which is a good idea when fishing in small lakes and ponds open to the public. The fish living in public water are often exposed to constant pressure. If you like going to large bodies of water, pressure does not affect the majority of fish. Plugs (opposite page) and flies (left) are effective bait for bluegill. Many anglers do not use artificial bait for bluegill for various reasons. Those who don't are missing out on some of the action, more action than they realize. The bluegill x green sunfish hybrid (left) grabbed an orange spider as soon as the fly landed. Nothing beats seeing the swirl of a fish smacking a lure on the surface. Bluegill are insect eaters, like trout, and a dream for anglers who tie their own flies, especially dry flies.

Males are darker in color than females.

THE AUTHOR

The first fish John Tertuliani caught was a bluegill. He has caught many more since. His childhood passion for the outdoors led to a Bachelor of Science in fisheries management and a Master of Science in biology. Even though his career of 24 years has allowed him to work with game species ranging from coho salmon to striped bass, John's passion for this tasty sunfish continues to grow.

catching Bluegill comes from a lifetime of fishing coupled with the knowledge that comes from a career in aquatic biology. Knowing where fish is the biggest challenge to making consistent catches. Bluegill cooperate most of the time. Catching bluegill and other species of sunfish will become easy if you follow the methods described in the book. Learn what a biologist knows about catching bluegill, proven methods no matter how you fish.

John is the author of *Smallmouth Bass and Streams: Thoughts on Flyfishing*. He can be reached at:

> Lotic Books
> P.O. Box 543
> Hilliard, Ohio 43026
> j.tertuliani@att.net

INDEX

A
Aberdeen hooks 55
active fish 99
approach 220
aquatic insects 66, 80
author 239
autumn 13

B
bait
  artificial 237
  crickets 39, 53, 217
  garden worms 39, 53, 59, 217
  leeches 217
  night crawlers 39, 59, 217, 231
  plastic bait 52, 57
  wax worms 39, 53, 58, 70, 217
bait hooks 37, 55
baitcasting tackle 28
baking 209
banks 87
barometer 171
batter 204
behavior 79
Berkley 52
bobber (float) stop 42, 46, 48, 71
bodies of water 221
bread 21, 39
breading 203

C
cane pole 24, 143
casting 34, 69, 72, 75
cattails 34
channel catfish 21
channels and troughs 86
circle hooks 37, 55, 58, 71, 148
cleaning fish 202
clippers 62, 235
color 7
comfort 151
coontail 66, 67
crayfish 66, 80
creek channel 65
crickets 53

D
daily activity 170
dam 65
daylight 9, 13
deep-frying 206
depth 66, 229
depth finder 191
density (water) 160
diving minnow baits 34
docks 93

E
eggs 10
epilimnion 161
equipment 235

F
feeding 12
fish
  active 99

fish—*continued*
  inactive 99
  suspended 98
fishing from shore 232
fishing with kids 139, 228
flats 84
flies
  ant 134
  behavior 124, 128
  dry 130, 132, 133
  fire fly 136
  phostrich 135
  poppers 137
  San Juan Worm 135
  spider 134
  terrestrial 132
  wet 131, 133
float stop 42, 46, 48, 71
floats 22, 31, 40, 42, 74, 148, 190
floating minnow baits 34
flooded shoreline 11
fly behavior 124, 221
fly line 105
  backing 117
  double taper 113
  leader 113
  loop connection 116
  multiple tip 111
  shooting head 111
  sinking 112
  sinking tip 110
  weight forward 109
forage (food) 80

G
garden worms 53
gravity 167
green sunfish 18

H
habitat 73, 82, 227
  winter 177
handling fish 197
heavy tackle 24, 29
hemostats 62
high water 230
hooks 37, 55, 58, 214
hybrid 19
hypolimnion 161

I
ice 179
  safety 187, 193
  tackle 181
image of your fly 221
in-line spinners 49
inactive fish 99
insects 80
invertebrate species 66

J
jigs 36, 51, 52, 189, 215

K
keep moving 227
knife 21, 62
knots 60, 69

L
leader 70
  fly 113

light tackle 24
lily 66
line 30, 105, 149, 212
  fly 105
littoral zone 64

live bait 39, 58, 59, 217, 237
location 150
lures 34, 215

M
maggots 172
matching the hatch 123
method 68
milfoil 66
minnow baits 34, 49
modified floats 54, 191
monofilament line 30, 69, 74, 188
movements 14
Mustad hooks 37, 55

N
natural bait 39
nests 10, 67
nymph 135

O
open water 83

P
patterns (fly) 131
photoperiod 10
plankton 80
plants 64, 65, 67, 73, 75, 219
plastic baits 52, 56, 57
plugs 34
points and humps 86
polarized glasses 75, 235
poles 24, 143
pondweed 66, 67
predators 8
preferred habitat 13
presentations 67, 100, 127
   horizontal 72
   vertical 69
pumpkinseed 19

R
redear sunfish 17, 151
reel 26, 117, 212
rigging for bait 218
rigging to fish over plants 220
rivers and streams 101
rock bass 20
rocks 80, 93, 227
rod 26, 27, 28, 71, 118, 211

S
sac fry 13
schooling behavior 8, 14
seasonal changes 162
setting the hook 71, 147
shellcracker 17
shore 232
sight fishing 140
sinkers 42, 45, 70
slip floats 42-48, 69, 180, 190
snapping turtle 21
soft plastic baits 52, 56, 57
spawning 9, 66, 68, 79
spin bubble 53, 74
spincasting tackle 26, 212
spinner baits 35, 49, 50, 216
spinning tackle 27, 212
split shot 41, 42, 70
spring 9
staging areas 65
strategy 163
stringers 198, 199
submerged plants 65
summer 11
sunfish species 17

sunlight 64, 79, 163, 225
suspended fish 98
swivel 45, 49, 53, 54, 70, 74

T
tackle 23
  baitcasting 28
  boxes 22, 51, 56, 234
  for children 142
  fillet knives 201
  floats 40, 42, 45, 74, 148, 190
  hooks 37, 55, 58, 214
  ice fishing 181
  in-line spinners 49
  jigs 36, 51, 215
  light tackle 24, 74
  line 30, 188, 212
  lures 34, 215
  heavy tackle 24, 29
  hooks 37, 55, 58, 214
  minnow baits 34, 49
  plastic baits 52, 56, 57
  plugs 34
  reel 26, 117, 212
  rod 26, 27, 28, 71, 118, 211
  sinkers 42, 45, 70
  spin bubble 53, 74
  spincasting 26, 212
  spinner baits 35, 49, 50, 216
  spinning 27, 212
  split shot 41, 42, 70
  storage 234
  swivel 45, 49, 53, 54, 70, 74
  tackle box 51, 56, 234
  treble hooks 35, 49
tactics 88
temperature 10, 13, 79, 160, 225
thermocline 161

tides 168
  inland effects 169
tools 62, 235
Trilene knot 60, 71
Tru-Turn hooks 37, 55

U

V
vegetation 65, 74, 177, 227
  fishing method 68

W
waves 65
water
  clarity 224
  layers 160
  public 69
  temperature 10, 13, 79, 160, 225
wax worms 39, 53, 58, 70, 217
weather 14
willow fly 80
wind 65, 226
winter 15
  finding fish 175
wire baskets 198, 199
wood 90 226
worms 39, 59, 217, 231